CAMBRIDGE LIBRARY COLLECTION

Books of enduring scholarly value

Earth Sciences

In the nineteenth century, geology emerged as a distinct academic discipline. It pointed the way towards the theory of evolution, as scientists including Gideon Mantell, Adam Sedgwick, Charles Lyell and Roderick Murchison began to use the evidence of minerals, rock formations and fossils to demonstrate that the earth was older by millions of years than the conventional, Bible-based wisdom had supposed. They argued convincingly that the climate, flora and fauna of the distant past could be deduced from geological evidence. Volcanic activity, the formation of mountains, and the action of glaciers and rivers, tides and ocean currents also became better understood. This series includes landmark publications by pioneers of the modern earth sciences, who advanced the scientific understanding of our planet and the processes by which it is constantly re-shaped.

A Bibliography of Indian Geology

Geologist and seismologist Richard Dixon Oldham (1858–1936) is best known for making two fundamental discoveries: in 1900 he identified primary, secondary and tertiary seismic waves, and in 1906 he presented evidence for the existence of the Earth's core. A fellow of the Royal Geographical Society, the Geological Society of London and the Royal Society, Oldham spent a large part of his career in Asia. Involved in the Geological Survey of India for twenty-five years, he played a key role in the development of geological research in the region (his revised *Manual of the Geology in India* is also reissued in this series). Originally published in 1888, the present work was the first bibliography of its kind and remains a valuable reference tool in earth sciences. Notably, Oldham chose to broaden the scope of the work by also listing papers on Indian mineralogy, geography, archaeology and botany.

Cambridge University Press has long been a pioneer in the reissuing of out-of-print titles from its own backlist, producing digital reprints of books that are still sought after by scholars and students but could not be reprinted economically using traditional technology. The Cambridge Library Collection extends this activity to a wider range of books which are still of importance to researchers and professionals, either for the source material they contain, or as landmarks in the history of their academic discipline.

Drawing from the world-renowned collections in the Cambridge University Library and other partner libraries, and guided by the advice of experts in each subject area, Cambridge University Press is using state-of-the-art scanning machines in its own Printing House to capture the content of each book selected for inclusion. The files are processed to give a consistently clear, crisp image, and the books finished to the high quality standard for which the Press is recognised around the world. The latest print-on-demand technology ensures that the books will remain available indefinitely, and that orders for single or multiple copies can quickly be supplied.

The Cambridge Library Collection brings back to life books of enduring scholarly value (including out-of-copyright works originally issued by other publishers) across a wide range of disciplines in the humanities and social sciences and in science and technology.

A Bibliography of
Indian Geology

Being a List of Books and Papers,
Relating to the Geology of British India
and Adjoining Countries,
Published Previous to the End of A.D. 1887

RICHARD DIXON OLDHAM

CAMBRIDGE
UNIVERSITY PRESS

CAMBRIDGE
UNIVERSITY PRESS

University Printing House, Cambridge, CB2 8BS, United Kingdom

Published in the United States of America by Cambridge University Press, New York

Cambridge University Press is part of the University of Cambridge.

It furthers the University's mission by disseminating knowledge in the pursuit of
education, learning and research at the highest international levels of excellence.

www.cambridge.org
Information on this title: www.cambridge.org/9781108064255

© in this compilation Cambridge University Press 2013

This edition first published 1888
This digitally printed version 2013

ISBN 978-1-108-06425-5 Paperback

A BIBLIOGRAPHY

OF

INDIAN GEOLOGY;

BEING

A LIST OF BOOKS AND PAPERS,

RELATING TO THE

GEOLOGY OF BRITISH INDIA AND ADJOINING COUNTRIES,

PUBLISHED PREVIOUS TO THE END OF A.D. 1887.

COMPILED BY

R. D. OLDHAM, A.R.S.M., F.G.S.,

DEPUTY SUPERINTENDENT OF THE GEOLOGICAL SURVEY OF INDIA.

PRELIMINARY ISSUE.

CALCUTTA:
PRINTED BY THE SUPERINTENDENT, GOVERNMENT PRINTING, INDIA.
1888.

NOTICE.

The author of the Bibliography having compiled the work entirely of his own taste and at his leisure, I cannot but express my thanks for his presentation of it to the Survey. Apart from the value of such a work to us, the aid it will surely afford to geologists in, and beyond, India makes its issue as one of the publications of the Survey a necessity also.

WILLIAM KING,
Director, Geological Survey of India.

PREFACE.

THE catalogue of books and papers contained in the following pages was commenced by me three years ago, and has been compiled in the spare time of a few short visits to Calcutta since then.

Under these circumstances uniformity or completeness would be impossible, but it is believed that the catalogue is now practically complete, except, perhaps, as regards palæontology, and that, while perhaps several papers may have been wrongly included, few, if any, of importance are omitted.

The geographical limits selected have been those of the territory subject to the Governor General and the States with which he has direct diplomatic relations; but it has been found advisable in some cases to overstep these limits and to include papers relating to countries outside them. In such cases only those papers have been included which have a more or less direct bearing on the geology of the territories with which this catalogue is directly concerned.

Besides purely geological papers there will be found many that deal principally with geography, archæology, botany, &c., which have been included on account of some isolated or scattered geological observations contained in them. In these cases it has often been difficult to determine whether a paper should be included or not; the general principle followed has been that where the notices are of early date, compared with our more complete knowledge of the region, they are included: but where of date subsequent to a tolerably complete account, and, consequently, wanting in fullness, originality, or accuracy, they have been excluded. Thus, the papers of this class will be found to be, for the most part, either of early date and dealing with districts now more or less well known, or modern and dealing with regions regarding which every scrap of information is still valuable. In applying this principle there is great room for individual differences of opinion, and it is possible that many papers have been wrongly admitted and others wrongly omitted.

Besides the papers catalogued there are many references to geology and economic mineralogy in the various gazetteers, district hand-books, settlement reports, &c., issued by Government. Where these are mere compilations at second or third hand by non-geological writers they have not been inserted, but where original or written by those who have a practical acquaintance with the subject they deal with they have been inserted.

The entries are arranged alphabetically according to the names of the authors, or, in the case of anonymous papers, under the name of the subject with which they deal. In order to ensure uniformity all names beginning with De, D', Du, Della, &c., O, Mac, Fitz, &c., have been indexed under the prefix, as this is in some cases invariably, in others frequently, an integral part of the word; in the case of the German prefix *von*, this has not been done. For the same reason compound names have always been indexed under the second half of the name, but, in the case of compound names and names with prefixes, cross references have been inserted in all cases where strict adherence to the rule would lead to delay or confusion. Papers of joint authorship are indexed according to that author's name which has alphabetical priority. They are, however, repeated, if necessary, after the papers of which the other author may be the sole writer.

Under each author's name the papers are arranged in chronological order, except that papers published in the same volume of the same periodical are arranged consecutively. Except where too long for complete reproduction, the author's title has been given in full; where too long the essential part has been given; and where it has been necessary to invent or amplify a title, the words added are inserted in square brackets; round brackets indicate part of the original title.

Following on the title come the references; of these the first is to the original publication in full of the paper; where more than one is given the second is usually to an abstract published in advance of publication in full, and will be recognised by its earlier date and greater brevity; following on this come reprints or notices arranged in chronological order. No reprint or notice has, however, been inserted except where it contains some substantial part of the original paper, or where the original is difficult of access. Where cross references are given they are printed in small capitals.

In abbreviating the names of periodicals, I have endeavoured to put them in such form that the periodical can be immediately recognised, and in those cases where the title is short and only referred to once or twice it has been printed in full. A list of the abbreviations used will be found at the end of this preface.

The manner of indicating volume, page, and date, is that in universal use and does not require explanation.

The limited time at my disposal and the deficiencies of the Calcutta libraries have prevented me from verifying every reference, but, as five-sixths at least have been extracted by myself, and a large number of the rest verified by comparison with the originals, or with the Royal Society's Catalogue, this will probably be of little importance.

In conclusion, I have to thank those numerous authors who have obliged me by correcting the lists of papers written by them and to express my gratitude in advance to all who will point out any errors or omissions in the catalogue.

LIST OF ABBREVIATED TITLES USED.

[Those marked with an asterisk are not in the library of the Geological Survey or of the Asiatic Society of Bengal : of some of the others the sets are incomplete.]

A

Abhandl. K. Akad. Wiss. Berlin.—Abhandlungen der Königlichen Akademie der Wissenschaften zu Berlin. 4°, Berlin.

Abhandl. K. Baier. Akad. Wiss.—Abhandlungen der Königlichen Baierischen Akademie der Wissenschaften zu München. 8°, Munich.

*****Abhandl. Naturhist. ges. Nürenberg.**—Abhandlungen der Naturhistorischen Gesellschaft zu Nürnberg. 8°, Nûrenberg.

Am. Assoc. Proc.—Proceedings of the American Association for the Advancement of Science. 8°, Salem.

Am. Jour. Sci.—The American Journal of Science and Art. 8°, New Haven.

Ann. de Chimie.—Annales de Chimie, ou Recueil de Mémoires concernant la Chimie et les Arts qui en dependent. 8°, Paris.

Ann. der Phys. u. Chem.—Annalen der Physik und Chemie : von J. C. Poggendorf. 8°, Leipzig.

Ann. des Mines.—Annales des Mines, ou Recueil de Mémoires sur l'exploitation des Mines, et sur les Sciences et les Arts qui s'y rapportent. 8°, Paris.

Ann. Mag. Nat. Hist.—Annals of Natural History, or Magazine of Zoology, Botany, and Geology. 8°, London.
——————————— The Annals and Magazine of Natural History, including. Zoology, Botany, and Geology. 8°, London.

*****Ann. Phil.**—Annals of Philosophy, or Magazine of Chemistry, Mineralogy, Mechanics, and the Arts. 8°, London.

Ann. Sci. Nat.—Annales des Sciences Naturelles, comprenant la Physiologie animale et Végétale, l'Anatomie comparée des deux regnes, la Zoologie, la Botanique, la Minéralogie, et la Géologie. 8° Paris.

As. Res.—Asiatick Researches, or Transactions of the Society instituted in Bengal for enquiring into the History and Antiquities, the Arts, Sciences, and Literature of Asia. 4°, Calcutta.

B

Bibl. Univ.—Bibliothèque Universelle des Sciences, Belles Lettres et Arts, faisant suite à la Bibliothèque Britannique rédigée à Genève. Partie des Sciences. 8°, Geneva.

Bol. Com. Map. Geol. Españ.—Boletin de la Comision del Mapa Geologico de España. Madrid.

Brit. Ass. Rep.—Reports of the British Association for the Advancement of Science. 8°, London.

Bull. Soc. Geogr. Paris.—Bulletin de la Société de Géographie de Paris. 8°, Paris.

Bull. Soc. Geol. Paris.—Bulletin de la Société Géologique de France. 8°, Paris.

C

Cal. Engin. Jour.—The Engineer's Journal and Railway Chronicle of India and the Colonies. 4°, Calcutta.

Cal. Jour. Nat. Hist.—The Calcutta Journal of Natural History. 8°, Calcutta.

Chem. News.—The Chemical News, a Journal of practical chemistry in all its applications to Pharmacy, Arts, and Manufactures. 8°, London.

D

Denk. K. K. Akad. Wiss. Wien.—Denk-schriften der K. K. Akademie der Wissen-schaften ; Mathematisch-Naturwissenschaftliche classe. 4°, Vienna.

E

*Edin. Jour. Sci.—The Edinburgh Journal of Science, exhibiting a view of the progress of discovery in Natural Philosophy, Chemistry, Mineralogy, Geology, Botany, &c. 8°, Edinburgh.

Edin. New Phil. Jour.—The Edinburgh New Philosophical Journal, exhibiting a view of the progressive improvements, &c., in the Sciences, &c. 8°, Edinburgh.

Edin. Phil. Jour.—The Edinburgh Philosophical Journal, exhibiting a view of the progress of discovery in Natural Philosophy, &c. 8°, Edinburgh.

F

*Froriep. Notizen.—Notizen aus dem Gebiete der Natur und Heilkunde. 4°, Erfurt und Weimar.

G

Geol. Mag.—The Geological Magazine or Monthly Journal of Geology. 8°, London.

Geol. Trans.—Transactions of the Geological Society of London. 4°, London.

Glean. Sci.—Gleanings in Science. 8°, Calcutta.

I

Ind. Ann. Med. Sci.—Indian Annals of Medical Science. 8°, Calcutta.

Ind. Economist.—The Indian Economist, a Monthly Journal devoted to Economic and Statistical Inquiries concerning India. 4°, Calcutta.

Ind. Forester.—The Indian Forester. Calcutta, afterwards Roorkee.

Ind. Jour. Arts. Sci.—Indian Journal of Arts, Sciences, and Manufactures. 8°, Madras.

Inland Customs Report.—Report on the Administration of the Imperial Customs Department. Folio, Calcutta.

Iron.—Iron, an Illustrated Weekly Journal of Science, Metals, and Manufactures in Iron and Steel. 4°, London.

J

Jahrb. K. K. Geol. Reichs. Wien.—Jahrbuch der K. K. Geologischen Reichsanstalt. 4°, Vienna.

Jour. Agri. Hort. Soc. Ind.—Journal of the Agricultural and Horticultural Society of India. 8°, Calcutta.

Jour. As. Soc. Beng.—The Journal of the Asiatic Society of Bengal. 8°, Calcutta.

Jour. Bo. As. Soc.—Journal of the Bombay Branch of the Royal Asiatic Society. 8°, Bombay.

Jour. Ceylon As. Soc.—Journal of the Ceylon Branch of the Royal Asiatic Society. 8°, Colombo.

*Jour. de Pharm.—Journal de Pharmacie et des Sciences Accessoires. 8°, Paris.

Jour. de Phys.—Journal de Physique, de Chimie, et de l'Histoire naturelle. 4°, Paris.

*Jour. des Mines.—Journal des Mines ou Recueil de Mémoires sur l'exploitation des Mines, et sur les Sciences et les Arts qui s'y rapportent. 8°, Paris.

Jour. Geol. Soc. Dublin.—Journal of the Geological Society of Dublin. 8°, Dublin.

Jour. Ind. Archip.—Journal of the Indian Archipelago and Eastern Asia. 8°, Singapore.

Jour. Roy. As. Soc.—Journal of the Royal Asiatic Society of Great Britain and Ireland. 8°, London.

Jour. Roy. Geog. Soc.—Journal of the Royal Geographical Society of London. 8°, London.

Jour. Roy. Geol. Soc. Dublin.—Journal of the Royal Geological Society of Ireland. 8°, Dublin.

Jour. Soc. Arts.—The Journal of the Society of Arts. 4°, London.

Jour. Straits, As. Soc.—Journal of the Straits Branch of the Royal Asiatic Society. 8°, Singapore.

L

*Lond. Jour. Bot.—London Journal of Botany and Kew Gardens Miscellany. 8°, London.

M

Mad. Jour. Lit. Sci.—Madras Journal of Literature and Science, published under the auspices of the Madras Literary Society and Auxiliary Royal Asiatic Society. 8°, Madras.

Mad. Mon. Jour. Med. Sci.—The Madras Monthly Journal of Medical Science. 8°, Madras.

Mad. Quart. Jour. Med. Sci.—Madras Quarterly Journal of Medical Science, 8°, Madras.

Mem. Am. Acad.—Memoirs of the American Academy of Arts and Sciences. 4°, Cambridge, U. S.

Mem. Geol. Surv. Ind.—Memoirs of the Geological Survey of India. 8°, Calcutta.

Mem. Wern. Soc. Edin.—Memoirs of the Wernerian Natural History Society. 8°, Edinburgh.

Mineral Mag.—The Mineralogical Magazine and Journal of the Mineralogical Society. 8°, London.

Min. Mitth.—Mineralogische Mittheilungen gesammelt von Gustav Tschermak. 4°, Vienna. (issued as part of Jahrb. K. K. Geol. Reichs. Wien).

Monatsber. K. Akad. Wiss. Berlin.—Monatsberichte der K. Preuss. Akademie der Wissenschaften zu Berlin. 8°, Berlin.

N

Nat Hist. Rev.—The Natural History Review and Quarterly Journal of Science. 8°, London and Dublin.

Nature.—Nature, a Weekly Illustrated Journal of Science. 4°, London.

Neu. Jahrb. Min. Geol.—Neues Jahrbuch für Mineralogie, Geognosie, Geologie und Petrefactenkunde. 8°, Stuttgart.

O

*Oesterr. Zeits. Berg. Huttenw.—Oesterreichische Zeitschrift fur Berg und Huttenwesen. 4°, Vienna.

P

Pal. Indica.—Memoirs of the Geological Survey of India, Palæontologia Indica, being Figures and Descriptions of the Organic remains procured during the progress of the Geological Survey of India. 4°, Calcutta.

Palæontographica.—Palæontographica : Beitrage zur Naturgeschichte der Vorzeit. 4°, Cassel.

Pal. Soc.—Palæontographical Society.

Petermann Mitth.—Mittheilungen aus Justus Perthes' Geographischer Anstalt über wichtige neue Erforschungen auf dem gesammt gebiete der Geographie. 4°, Gotha.

*Pharmaceut. Jour.—Pharmaceutical Journal and Transactions. 8°, London.

Phil. Mag.—The Philosophical Magazine or Annals of Chemistry, Mathematics, Astronomy, Natural History, and General Science. 8°, London.

 The London, Edinburgh, and Dublin Philosophical Magazine and Journal of General Science.

Phil. Trans.—Philosophical Transactions of the Royal Society of London. 4°, London

Proc. As. Soc. Beng.—Proceedings of the Asiatic Society of Bengal. 8°, Calcutta.

Proc. Geol. Soc.—Proceedings of the Geological Society of London. 8°, London.

Proc. Inst. Civ. Eng.—Proceedings of the Institution of Civil Engineers. 8°, London.

Proc. Liverpool Geol. Soc.—Proceedings of the Liverpool Geological Society. 8°, Liverpool.

Proc. Liverpool Lit. Phil. Soc.—Report of the Proceedings of the Literary and Philosophical Society of Liverpool. 8°, Liverpool.

Proc. Boston Soc. Nat. Hist.—Proceedings of the Boston Society of Natural History. 8°, Boston.

Proc. Roy. Dublin Soc.—Scientific Proceedings of the Royal Dublin Society. 8°, Dublin.

Proc. Roy. Geog. Soc.—Proceedings of the Royal Geographical Society and Monthly Record of Geography. 8°, London.

Proc. Roy. Irish. Acad.—Proceedings of the Royal Irish Academy. 8°, Dublin.

Proc. Roy. Soc.—Proceedings of the Royal Society of London. 8°, London.

Proc. Zool. Soc.—Proceedings of the Zoological Society of London. 8°, London.

Prof. Pap. Ind. Eng.—Professional Papers on Indian Engineering. 8°, Roorkee.

Prof. Pap. Thomason Engin. Coll.—Professional Papers printed at the Civil Engineering College, Roorkee. 8°, Roorkee.

Q

Quart. Jour. Geol. Soc.—Quarterly Journal of the Geological Society of London. 8°, London.

Quart. Jour. Sci.—Quarterly Journal of Science, Literature, and Arts. 8°, London.

R

* Rec. Gen. Sci. Records of General Science. 8°, London.

Rec. Geol. Surv. Ind.—Records of the Geological Survey of India. 8°, Calcutta.

Rep. Gov. Mus. Mad.—Reports on the Government Central Museum, Madras, and on the Government Museums at Bellary, Coimbatore, Cuddalore, Mangalore, Ootacamund, Rajahmundry. 8°, Madras.

Rep. Surveyor General Ind.—General Report on the operations of the Survey of India Department, administered under the Government of India. Folio, Calcutta.

S

Scottish Geog. Mag.—The Scottish Geographical Magazine. 8°, Edinburgh.

Sel. Pub. Corr. N.-W. P.—Selections from the Public Correspondence of the Government of the North-Western Provinces. 8°, Allahabad.

Sel. Pub. Corr. Punjab.—Selections from the Public Correspondence of the Punjab Government. 8°, Lahore.

Sel. Rec. Beng. Gov.—Selections from the Records of the Bengal Government. 8°, Calcutta.

Sel. Rec. Bo. Gov.—Selections from the Records of the Bombay Government. 8°, Bombay.

Sel. Rec. Gov. Ind.—Selections from the Records of the Government of India. 8°, Calcutta.

Sel. Rec. Gov. N.-W. P.—Selections from the Records of the Government of the North-Western Provinces. 8°, Allahabad.

Sel. Rec. Mad. Gov.—Selections from the Records of the Madras Government. 8°, Madras.

Sel. Rec. Punjab Gov.—Selections from the Records of the Government of the Punjab and its dependencies. 8°, Lahore.

Sitz. K. Akad. Wien.—Sitzungsberichte der Mathematisch-naturwissenschaftlichen classe der Kaiserlichen Akademie der Wissenschaften. 8°, Vienna.

Sitz. K. Baier. Akad. Wiss.—Sitzungsberichte der Königlichen Baierischen Akademie, der Wissenschaften zu München. 8°, Munich.

*****Sitz. K. Böhm. Gesel. Wiss.**—Sitzungsberichte der Königl Böhmischen gesellschaft der Wissenschaften in Prag. 8°, Prague.

T

Trans. Am. Phil. Soc.—Transactions of the American Philosophical Society, held at Philadelphia, for promoting useful knowledge. 4°, Philadelphia.

Trans. Bo. Geog. Soc.—Transactious of the Bombay Geographical Society. 8°, Bombay.

Trans. Edin. Geol. Soc.—Transactions of the Edinburgh Geological Society. 8°, Edinburgh.

Trans. Geol. Soc. Cornwall.—Transactions of the Royal Geological Society of Cornwall. 8°, Penzance.

Trans. Lit. Soc. Bombay.—Transactions of the Literary Society of Bombay. 4°, London.

Trans. Med. Phys. Soc. Bombay.—Transactions of the Medical and Physical Society of Bombay. 8°, Bombay.

Trans. Med. Phys. Soc. Calcutta.—Transactions of the Medical and Physical Society of Calcutta. 8°, Calcutta.

Trans. N. Eng. Inst. Min. Mech. Eng.—Transactions of the North of England Institute of Mining and Mechanical Engineers. 8°, Newcastle-upon-Tyne.

Trans. Roy. As. Soc.—Transactions of the Royal Asiatic Society of Great Britain and Ireland. 4°, London.

Trans. Roy. Soc. Edin.—Transactions of the Royal Society of Edinburgh. 4°, Edinburgh.

V

Verhand. Batav. Genootsch.—Verhandelingen van het Bataviaasch genootschap der Kunsten en Wetenschappen. 4°, Batavia.

Verhandl. K. K. Geol. Reichs. Wien.—Verhandlungen der K. K. Geologischen Reichsanstalt. 4°, Vienna.

Vierteljahrs. Naturf. Ges. Zurich.—Vierteljahrsschrift der Naturforschenden Gesellschaft in Zurich. 8°, Zurich.

Z

Zeits. Allg. Erdkund.—Zeitschrift für allgemeine Erdkunde. 8°, Berlin.

Zeits. Deutsch. Geol. Ges.—Zeitschrift der Deutschen Geologischen Gesellschaft. 8°, Berlin.

Zeits. gesammt. Naturw. Halle.—Zeitschrift für die gesammten Naturwissenschaften; heraus-gegeben von dem Naturwissenschaftlichen Vereine fur Sachsen und Thüringen in Halle. 8°, Berlin.

A BIBLIOGRAPHY

OF

INDIAN GEOLOGY.

A

Abbot, J.

1. An account of a Remarkable Aerolite which fell at the village at Manic-gaon, near Eidulabad, in Khandeesh: *Jour. As. Soc. Beng.*, XIII, 880—886, (1844).

2. On Kunker Formations, with specimens: *Jour. As. Soc. Beng.*, XIV, 442—444, (1845).

3. Account of certain Agate Splinters found in the clay stratum bordering the River Narbudda: *Jour. As. Soc. Beng.*, XIV, 756—758, (1845).

4. Remarks upon the occurrence of granite in the bed of the Narbudda: *Jour. As. Soc. Beng.*, XIV, 821-822, (1845).

5. Account of the process employed for obtaining Gold from the Sand of the River Beyass; with a short account of the Gold Mines of Siberia: *Jour. As. Soc. Beng.*, XVI, 266—271, (1847).

6. Extracts from a letter, descriptive of Geological and Mineralogical Observations in the Huzaree district, dated Camp Puhli, in Huzaree, 19th June 1847: *Jour. As. Soc. Beng.*, XVI, 1135—1140, (1847).

7. Inundation of the Indus, taken from the lips of an eye-witness (Ushruff Khan), A.D. 1842: *Jour. As. Soc. Beng.*, XVII, 230—232, (1848).

Adam, J.

1. On the Geology of the banks of the Ganges from Calcutta to Cawnpore: *Geol. Trans.*, 1st series, V, 346—352, (1821).

2. Geological notices and miscellaneous remarks relative to the district between the Jumna and Nerbuddah; with an Appendix containing an account of the rocks found in the Baitool Valley, in Berar, and on the hills of the Gundwana Range; together with remarks made on a march from Hussingabad to Saugar, and from thence to the Ganges: *Mem. Wern. Soc. Edin.*, IV, 24—57, (1822).

3. Account of Barren Island in the Bay of Bengal: *Jour. As. Soc. Beng.*, I, 128—131, (1832).

4. Memoranda on the Geology of Bundelcund and Jubbulpore: *Jour. As. Soc. Beng.*, XI, 392—410, (1842).

Adams, A. Leith.

1. Wanderings of a Naturalist in India, the Western Himalayas, and Cashmere: 8°. Edinburgh, 1867.

B

Adams, A. Leith,—cont.

2. Has the Asiatic Elephant been found in a fossil state ? With additional remarks by G. Busk: *Quart. Jour. Geol. Soc.*, XXIV, 496—499, (1868).

Agha Abbas.

1. Journal of a Tour through parts of the Punjab and Afghanistan, in the year 1837. Arranged and translated by Major R. Leech, by whom the tour was planned and instructions furnished : *Jour. As. Soc. Beng.*, XII, 564—621, (1843).

Aitchison, J. E. T.

1. On the vegetation of the Jhelum district of the Punjab : *Jour. As. Soc. Beng.*, XXXIII, 290—320. [Contains some geological notes at p. 291.]

Alexander, J. E.

1. Notice in regard to the Saline Lake of Loonar, situated in Berar, East Indies : *Edin. Phil. Jour.*, XI, 308—311, (1824).

Allardyce, J.

1. On the granitic formation and direction of the Primary Mountain chains of Southern India : *Mad. Jour. Lit. Sci.*, IV, 327—336, (1836).

Allen, C. L.

1. On the composition of two specimens of Jade (Karakash) : *Chem. News*, XLVI, 216, (1882).

Alum.

1. Alum works in Kutch : *Glean. Sci.*, III, 384—385, (1831).

Amato, *Père,* **Gieuseppe d',** *see* D'AMATO.

Andaman Committee.

1. Report of the Andaman Committee : *Sel. Rec. Gov. Ind.*, XXV, 4—28, (1859).

Anderson, James.

1. Account of the strata at the Diamond mines of Malivully : *Edin. Phil. Jour.*, III, 72—73, (1820).

Anderson, John.

1. A report on the expedition to Western Yunan *via* Bhamô. 8°. Calcutta, 1871.

2. Mandalay to Momien : A narrative of the two expeditions to Western China, of 1868 and 1875, under Col. Edward B. Sladen and Col. Horace Brown. 8°. London, 1876.

Andresen, T. F.

1. [Copper mines in Alwar] : *Mining Journal*, 1884, p. 1029.

Ansted, D. T.

1. Notice of the Coal of India, being an Analysis of a Report communicated to the Indian Government on this subject : *Brit. Ass. Rep.*, 1846, pt. ii, pp. 63—65.

Antimony.

1. Antimony mines in the Punjab, [Lahoul], reprinted from the *Madras Spectator*, 19th March 1857: *Mad. Jour. Lit. Sci.*, XVII, (new series I), 254—257, (1857).

Applegath, F.

1. On the Geology of a part of the Masulipatam District (abridged) : *Quart. Jour. Geol. Soc.*, XIX, 32-35, (1863).

1. [On the supposed Kistna coal]: *Jour. Soc. Arts*, XXX, 590, (1882).

Arracan.

1. Appearance of a new Volcanic Island off the coast of Arracan : *Cal. Jour. Nat. Hist.*, IV, 455, (1844). See also CHEDUBA.

Assam.

1. Mineral Indigo [vivianite] from Assam : *Cal. Jour. Nat. Hist.*, III, 153, (1843).

Atkinson, E. T.

1. The Himalayan Districts of the North-Western Provinces: *Gazetteer, N.-W. Provinces*, Vol. X, Allahabad, 1882, [Physical Geography, pp. 61 —110 ; Economic Mineralogy, pp. 259—298).

Austen, H. H. Godwin.

1. On the lacustrine, or Karéwah, deposits of Kashmere : *Quart. Jour. Geol. Soc.*, XV, 221—229, (1859).

2. Notes on the valley of Kashmir : *Jour. Roy. Geog. Soc.*, XXXI, 30—37, (1859).

3. On the Geology of part of the North-Western Himalayas ; with Notes on the fossils, by Messrs. T. Davidson, R. Etheridge, & S. P. Woodward : *Quart. Jour. Geol. Soc.*, XX, 383—387, (1864), XXII, 34, (1866).

4. Notes on the Sandstone formation, &c., near Buxa Fort, Bhootan Dooars : *Jour. As. Soc. Beng.*, XXXIV, pt. ii, 106—107, (1865) ; *Proc. As. Soc. Beng.*, '1865, pp. 90—91.

5. On the Carboniferous rocks of the valley of Cashmere, with Notes on the Carboniferous Brachiopoda, by T. Davidson, and an Introduction and résumé, by R. A. S. Godwin-Austen : *Quart. Jour. Geol. Soc.*, XXI, 492, (1865) ; XXII, 29—35, (1866).

6. Notes on the Pangong-lake district of Ladakh, from a journal made in 1863 : *Jour. Roy. Geog. Soc.*, XXXVII, 343—363, (1867) ; *Jour. As. Soc. Beng.*, XXXVII, pt. ii, 84—117, (1868) ; *Sel. Rec. Gov. Ind.*, LXXI, 1—24, (1869).

7. Notes on Geological features of the country near the foot of the hills in the Western Bhootan Dooars : *Jour. As. Soc. Beng.*, XXXVII, pt. ii, 117—123, (1868).

8. Extract from narrative report of Captain H. H. Godwin-Austen, in charge of No. 6 Topographical Party, Cossyah and Garrow Hills Survey: *Sel. Rec. Gov. Ind.*, LXXI, 140—147, (1869).

9. Memorandum as to the Geology of the Jaintia hills and as to the distri bution of the tribes : *Sel. Rec. Gov. Ind.*, LXXIV, 69—73, (1869).

Austen, H. H. Godwin,—cont.

10. Notes to accompany a Geological map of a portion of the Khasi hills near Longitude 91° E : *Jour. As. Soc.Beng.*, XXXVIII, pt. ii, 1—27, (1869).

11. Notes from Assaloo, North Cachar, on the great earthquake, January 10th, 1869 : *Proc. As. Soc. Beng.*, 1869, pp. 91—99.

12. Earthquake in the Cachar hills. Extracts from letters : *Jour. Roy. Geog. Soc.*, XIII, 370—372, (1869).

13. Notes on the Geology and Physical features of the Jaintia hills : *Jour. As. Soc. Beng.*, XXXVIII, pt. ii, 151—156, (1869) ; *Proc. As. Soc. Beng.*, 1869, pp. 64—65.

14. On the Garo hills : *Jour. Roy. Geog. Soc.*, XLIII, 1—46, (1873). [Geological Appendix, pp. 42—46.]

15. [Geology of parts of Naga hills and Manipur] : *Rep. Surveyor General Ind.*, 1872—73, pp. 79—84, (1874).

16. Notes on the Geology of part of the Dafla hills, Assam ; lately visited by the Force under Brigadier-General Stafford : *Jour. As. Soc. Beng.*, XLIV, pt. ii, 35—41, (1875).

17. Exhibition of a Celt found at Shillong : *Proc. As. Soc. Beng.*, 1875, p. 158.

18. The evidence of a past Glacial action in the Nágá hills, Assam : *Jour. As. Soc. Beng.*, XLIV, pt. ii, 209—213, (1875).

19. Remarks on Himalayan glaciation : *Proc. As. Soc. Beng.*, 1877, p. 4.

20. On the post-tertiary and more recent Deposits of Cashmir and the Upper Indus Valley : *Brit. Ass. Rep.*, 1880, p. 589.

21. Presidential address to Section E.: *Brit. Ass. Rep.*, 1883, pp. 576—589. *Proc. Roy. Geog. Soc.*, new series, V, 610—625; *Nature*, XXVIII, 552—558, (1883).

Aytoun, A.

1. Report on the Geological Survey of the Belgaum collectorate : *Trans. Bo. Geog. Soc.*, XI, 1—16, 30—60, (1854); WESTERN INDIA, pp. 378—397, (1857).

2. Geology of the Southern Concan: *Edin. New Phil. Jour.*, 2nd series, IV, 67—85, (1856).

3. On the Origin and Distribution of the Regur, or Black Cotton, soils of the Indian Peninsula. 8°, Edinburgh. Printed for Private circulation, 1863.

B

Babington, B.

1. Remarks on the Geology of the country between Tellicherry and Madras : *Geol. Trans.*, 1st series, V, 328—329, (1821).

Babington, C. L.

1. [Iron of Kutterbagga, 20 miles N.-E. of Sumbulpore] : *Jour. As. Soc. Beng.*, XII, 163—164, (1843).

Babington, S.

1. On the Island of Salsette : *Geol. Trans.*, 1st series, V, 1—3, (1821).

Baden Powell, B. H., *see* POWELL, B. H. BADEN.

Baily, W. H.
  1. On Tertiary fossils of India : *Brit. Ass. Rep.*, 1859, pp. 97—98.

Baird Smith, R., *see* SMITH, R. BAIRD.

Baker, W. E.
  1. Description of the Fossil Elephant's Tooth from Somrotee, near Nahun : *Jour. As. Soc. Beng.*, III, 638, (1834).
  2. On the Fossil Elk of the Himalaya : *Jour. As. Soc. Beng.*, IV, 506, (1835).
  3. Selected specimens of the Sub-Himalayan Fossils in the Dádúpur collection : *Jour. As. Soc. Beng.*, IV, 565—570, (1835).
  4. Note on the Fossil Camel of the Sub-Himalayas : *Jour. As. Soc. Beng.*, IV, 694—695, (1835).
  5. Report on a line of Levels taken by order of the Right Honourable the Governor General, between the Jumna and Sutlej rivers : *Jour. As. Soc. Beng.*, IX, 688—693, (1840).
  6. Note on a Fossil Antelope from the Dadoopoor Museum : *Jour. As. Soc. Beng.*, XII, 769—770, (1843).
  7. Remarks on the Alla Bund and on the drainage of the Eastern part of the Scinde Basin : *Trans. Bo. Geog. Soc.*, VII, 186—188, (1846).
  8. Memorandum on the prospect of remuneration in working the Iron Mines of the Raneegunge district. With a report by Professor Oldham : *Jour. As. Soc. Beng.*, XXII, 484—491, (1853).
  9. Report on the upper portion of the Eastern Narra, its sources of supply, and the feasibility of restoring it as a permanent stream : *Sel. Rec. Bo. Gov.*, XLV, 1—5, (1857).

Baker, W. E. *and* **Durand, H. M.**
  1. Table of Sub-Himalayan Fossil genera in the Dádúpur|Collection : *Jour. As. Soc. Beng.*, V, 291-293, 486-504, 661—669, 739-740, (1836). Fossil monkey's jaw in *An. Sci. Nat. Paris*, (*Zoöl.*), VII, 370—372, (1837) ; *Phil. Mag.*, XI, 33—36, (1837) ; *Edin. New Phil. Jour.*, XXIII, 216-217, (1837).
  2. Fossil Remains of the smaller carnivora from the Sub-Himalayas : *Jour. As. Soc. Beng.*, V, 576—584, (1836).

Balfour, E.
  1. Marbles (and Limestones) of the Madras Presidency : *Sel. Rec. Mad. Gov.*, 1st ser., II, 5—41, (1854).
  2. On the Iron ores, the Manufacture of Iron and Steel, and the Coals of the Madras Presidency. 8°. Madras, (1855).
  3. Report on the Government Central Museum : *Sel. Rec. Mad. Gov.*, 2nd ser., XXXIX, (1857).
  4. Cyclopædia of India and of Eastern and Southern Asia, commercial, industrial, and scientific : 8°. Madras, 1887. *2nd edition*, 5 vols., 8°, Madras, 1871—73. *3rd edition*, 3 vols., 8°, London, 1885.
  5. Index to Geological Papers in the Madras Journal of Literature and Science : *Mad. Jour. Lit. Sci.*, XXI (new ser., V), 158—164, (1859).

Ball, V.

1. Stone implements in Bengal: *Proc. As. Soc. Beng.*, 1865, pp. 127-128.

2. [On chipped implements from Bengal]: *Proc. As. Soc. Beng.*, 1867, p. 143.

3. List of localities in India where ancient stone implements have been discovered: *Proc. As. Soc. Beng.*, 1867, pp. 147—153.

4. The Ramgarh coal-field: *Mem. Geol. Surv. Ind.*, VI, 109—135, (1867).

5. On stone implements: *Proc. As. Soc. Beng.*, 1868, p. 177.

6. On the ancient Copper Miners of Singhbúm: *Proc. As. Soc. Beng.*, 1869, pp. 170—175.

7. On the occurrence of gold in the district of Singhbhúm, &c.: *Rec. Geol. Surv. Ind.*, II, 11—14, (1869).

8. Brief notes on the Geology and on the Fauna in the neighbourhood of Nancowry Harbour, Nicobar Islands: *Jour. As. Soc. Beng.*, pt. ii, XXXIX, 25—34, (1870); *Proc. As. Soc. Beng.*, 1869, pp. 250—253.

9. Notes on the Geology of the vicinity of Port Blair, Andaman Islands: *Jour. As. Soc. Beng.*, pt. ii, XXXIX, 231—239, (1870).

10. Remarks on Celts found in Singhbhúm: *Proc. As. Soc. Beng.*, 1870, p. 268.

11. On the occurrence of argentiferous galena and copper in the district of Manbhúm, south-west frontier of Bengal: *Rec. Geol. Surv. Ind.*, III, 74—76, (1870).

12. On the copper deposits of Dhalbhúm and Singhbhúm: *Rec. Geol. Surv. Ind.*, III, 94—103, (1870).

13. The Raigur and Hingir (Gangpúr) coal-field: *Rec. Geol Surv. Ind.*, IV, 101—107, (1871).

14. The Chopé coal-field: *Mem. Geol. Surv. Ind.*, VIII, 347—352, (1872).

15. The Bisrampúr coal-field: *Rec. Geol. Surv. Ind.*, VI, 25—41, (1873).

16. Barren Island and Narkondám: *Rec. Geol. Surv. Ind.*, VI, 81—90, (1873); *Geol. Mag.*, 2nd Decade, VI, 16—27, (1879).

17. On the discovery of a new locality for copper in the Nárbada Valley: *Rec. Geol. Surv. Ind.*, VII, 62—63, (1874).

18. On the building and ornamental stones of India: *Rec. Geol. Surv. Ind.*, VII, 98—122, (1874).

19. Geological notes made on a visit to the coal recently discovered in the country of the Luni Patháns, south-east corner of Afghanistan: *Rec. Geol. Surv. Ind.*, VII, 145—158, (1874).

20. The Raigarh and Hingir coal-field: *Rec. Geol. Surv. Ind.*, VIII, 102—121, (1875).

21. On an ancient Kitchen Midden at Chaudwar, near Cuttack: *Proc. As. Soc. Beng.*, 1876, pp. 120-121.

22. On stone implements found in the Tributary States of Orissa: *Proc. As. Soc. Beng.*, 1876, pp. 122-123.

23. On the Atgarh sand-stones near Cuttack: *Rec. Geol. Surv. Ind.*, X, 63—68, (1877).

24. On the Geology of the Máhanadi Basin and its vicinity: *Rec. Geol. Surv. Ind.*, X, 167—186, (1877).

Ball, V.,—cont.

25. On the Diamonds, gold and lead ores of the Sambalpur district: *Rec. Geol. Surv. Ind.*, X, 186—192, (1877).

26. Geology of the Rajmehal Hills: *Mem. Geol. Surv. Ind.*, XIII, 155—248, (1877).

27. Remarks on the Abstract and Discussion of Dr. O. Feistmantel's paper entitled "Giant Kettles (pot holes) caused by water-action in Streams in the Rajmahal Hills and the Barakur district": *Proc. As. Soc. Beng.*, 1877, pp. 140—143.

28. On the origin of the Kumaon lakes: *Rec. Geol. Surv. Ind.*, XI, 174—182, (1878).

29. On the Aurunga and Hutar coal-fields, and the iron ores of Palamow and Toree: *Mem. Geol. Surv. Ind.*, XV, pt. i, 1—27, (1878).

30. On the New Geological Map of India: *Brit. Ass. Rep.*, 1878, pp. 532-533.

31. Exhibition of two stone implements from Parisnath Hill: *Proc. As. Soc. Beng.*, 1878, p. 125.

32. On Stilbite from veins in metamorphic (gneiss) rocks in Western Bengal: *Jour. Roy. Geol. Soc. Dublin*, V, 114-115, (1879).

33. On spheroidal jointing in metamorphic rocks in India and elsewhere producing a structure resembling glacial "roches moutonnées": *Proc. Roy. Dub. Soc.*, new series, II, 341—346, (1879); *Jour. Roy. Geol. Soc. Dublin*, V, 193—198, (1879).

34. On the evidence in favour of the belief in the existence of floating ice in India during the deposition of the Talchir (Permian, or Permo-triassic) rocks: *Proc. Roy. Dub. Soc.*, new series, II, 430—436, (1879); *Jour. Roy. Geol. Soc. Dublin*, V, 223—229, (1879).

35. On the coal-fields and Coal Production of India: *Proc. Roy. Dub. Soc.*, new series, II, 496—523, (1879); *Jour. Roy. Geol. Soc. Dublin*, V, 230—257, (1879); *Brit. Ass. Rep.* 1879, 334—335.

36. On the mode of occurrence and distribution of Gold in India: *Proc. Roy. Dub. Soc.*, new series, II, 524—546, (1880); *Jour. Roy. Geol. Soc. Dublin*, V, 258—280, (1879).

37. On the mode of occurrence and distribution of Diamonds in India: *Proc. Roy. Dub. Soc.*, new series, II, 551—589, (1880); *Jour. Roy. Geol. Soc., Dublin*, VI, 10—48, (1880).

38. On the forms and geographical distribution of ancient Stone Implements in India: *Proc. Roy. Irish Acad.* (Pol., Lit. & Antiqs.), 2nd series, I, 388—414, (1879); *Brit. Ass. Rep.*, 1878, pt. ii, p. 394.

39. Jungle life in India or the journeys and journals of an Indian Geologist: 8°. London, 1880.

40. The Diamonds, Coal, and Gold of India; their mode of occurrence and distribution. 8°. London, 1881.

41. On the identification of certain localities mentioned in my paper on the Diamonds in India: *Jour. Roy. Dub. Soc.*, VI, 69—70, (1881).

42. Geology of the Districts of Manbhúm and Singhbhúm: *Mem. Geol. Surv. Ind.*, XVIII, 61—150, (1881).

43. Exhibition of an ancient stone implement made of magnetic iron ore: *Proc. As. Soc. Beng.*, 1881, 120.

Ball, V.,—cont.

44. On the origin of the so-called Kharakpur Meteorite: *Proc. As. Soc. Beng.*, 1881, pp. 140-142.

45. On the identification of certain Diamond Mines in India, which were known to, and worked by, the ancients, especially those which were visited by Tavernier. With a Note on the history of the Koh-i-nur: *Jour. As. Soc. Beng.*, L, pt. ii, 31—44, (1881).

46. Additional note on the identification of the ancient Diamond Mines visited by Tavernier: *Jour. As. Soc. Beng.*, L, pt. ii, 219—223, (1881).

47. A Manual of the Geology of India. Part III, Economic Geology. 8°. Calcutta, 1881.

48. On the coal-bearing rocks of the Upper Rer and Mand valleys in Western Chota Nagpur: *Rec. Geol. Surv. Ind.*, XV, 108—121, (1882).

49. The Mineral Resources of India and their development: *Jour. Soc. Arts*, XXX, 577—590, (1882).

50. On the mode of occurrence of precious stones and metals in India: *Brit. Ass. Rep.*, 1884, pp. 731—732.

51. On some effects produced by Landslips and movements of the soil-cap which are generally attributed to other agencies: *Jour. Roy. Geol. Soc. Dublin*, VI, 193—200, (1885).

52. On some recent additions to our knowledge of the gold-bearing rocks of Southern India: *Jour. Roy. Geol. Soc. Dublin*, VI, 201—206, (1885).

53. A geologist's contribution to the History of Ancient India: *Jour. Roy. Geol. Soc. Dublin*, VI, 221—263, (1885).

54. On the newly discovered Sapphire Mines in the Himalayas: *Proc. Roy. Soc. Dublin*, IV, 393—395, (1885).

55. The mineral resources of India and Burmah, being a lecture delivered at the Colonial and Indian Exhibition on the 5th June, 1886: *Mining Journal*, LVI, 674—675, (1886).

56. Zinc and zinc-ores: their mode of occurrence, metallurgy and History in India, with a Glossary of oriental and other titles used for zinc, its ores, and alloys: *Proc. Roy. Dub. Soc.*, V, 321—331, (1887).

Banerji, Chandrasekhara.
1. The Kaimur range: *Jour. As. Soc. Beng.*, XLVI, pt. i, 16—36, (1877).

Barratt, J.
1. On the Beerbhoom Iron Works. 8°. Calcutta, 1856, (printed privately).

2. Report on the Survey of the mineral deposits in Kumaon: *Sel. Rec. Gov. Ind.*, XVII, 62—81, (1856).

3. On Carbonates of Alumina, sesquioxide of Chromium and Iron: *Chemical News*, I, 110—111, (1860).

Barren Island.
1. [Report on Barren Island]: *Proc. As. Soc. Beng.*, 1866, pp. 212—217.

Batten, J. H.
1. Note of a visit to the Niti pass of the grand Himalayan chain: *Jour. As. Soc. Beng.*, VII, 310—316, (1838).

Batten, J. H.,—cont.

2. A few notes on the subject of the Kumaon and Rohilcund Turaie: *Jour. As. Soc. Beng.*, XIII, 877—914, (1844) ; Kumaon, pp. 128—146, (1878).

Batten, J. H., *and* **Herbert, J. D.**

1. Journal of Captain Herbert's tour from Almorah in a N.-W., W., and S. W. direction, through parts of the province of Kumaon and British Gurhwal, chiefly in the centre of the hills, *vide* No. 66, Indian Atlas: *Jour. As. Soc. Beng.*, XIII, 734—764, (1844).

Batten, J. H., *and* **Manson, E.**

1. Journal of a visit [by E. Manson] to the Melum and the Oonta Dhoora pass in Juwahir: *Jour. As. Soc. Beng.*, XI, 1157—1181, (1842).

Bauerman, H.

1. Report on the Iron ores of India: *Gazette of India Supplement*, 1874, pp. 1457—1459, 1494—1496.

Becher, J.

1. Letter [on the cataclysm of the Indus, 18th August, 1858] : *Jour. As. Soc. Ben.*, XXVIII, 219—228, 302, (1859).

Beckett, J. O'B.

1. Report on the Iron mines of Puttee Gowar in Zillah Kumaon and on the iron and copper mines of Puttee Lobah in Zillah Gurhwal : *Sel. Rec. Gov. N.-W. P.*, III, 67—76, (1853), and new series, III, 22—38, (1867).

Bedford, H.

1. Extract from the Journal of Apothecary H. Bedford, deputed to Yenang-Young, in Ava, in search of Fossil Remains : *Glean. Sci.*, III, 168—170, (1831).

Bell, H.

1. Masonry in Trap country, [Traps of Western India]: *Prof. Pap. Ind. Eng.*, 2nd series, I, 162—172, (1872).

Bell, H. C. P.

1. The Maldive Islands : an account of the Physical features, climate, history, &c. Flscp. Colombo, 1883.

Bell, T. L.

1. On the geology of the neighbourhood of Kotah, Deccan: *Quart. Jour. Geol. Soc.*, VIII, 230—233, (1852) ; Western India, 303—307, (1857).

2. Further account of a boring at Kotah, Deccan; and a notice of an Ichthyolite from that place : *Quart. Jour. Geol. Soc.*, IX, 351—352, (1853).

Bellew, H. W.

1. Journal of a political mission to Afghanistan in 1857, under Major (now Colonel) Lumsden ; with an account of the country and people. 8°. London, 1862.

2. Record of the March of the mission to Seistan. 8°. Calcutta, 1873.

3. Kashmir and Kashgar : a narrative of the journey of the Embassy to Kashgar in 1873—74. 8°. London, 1875.

Benson.

1. Fossil bones from Jubbulpore: *Jour. As. Soc. Beng.*, II, 151, (1833).

Benza, P. M.

1. Geological sketch of the Nilgherries: *Jour. As. Soc. Beng.*, IV, 413-437, (1835).

2. Notes on the geology of the country between Madras and the Neilgherry hills, *via* Bangalore and *via* Salem: *Mad. Jour. Lit. Sci.*, IV, 1—27, (1836).

3. Memoir on the geology of the Neilgherry and Koondah mountains: *Mad. Jour. Lit. Sci.*, IV, 241—299, (1836).

4. Notes, chiefly geological, of a journey through the Northern Circars, in the year 1835: *Mad. Jour. Lit. Sci.*, V, 43—71, (1837); part in *Vizagapatam Manual.* 8°. Madras, 1869, pp. 29—36.

Bettington, Albemarle.

1. Memorandum on certain fossils, more particularly a new Ruminant found at the Island of Perim in the gulf of Cambay: *Jour. Roy. As. Soc.*, VIII, 340—348, (1846).

Betts, C.

1. Hot spring at Pachete [Damuda valley]: *Jour. As. Soc. Beng.*, II, 46, (1833).

Beveridge, H.

1. Were the Sundarbans inhabited in ancient times? *Jour. As. Soc. Beng.*, XLV, pt. i, 71—76, (1876).

Beyrich.

1. Ueber einige Trias ammoniten aus Asien: *Monatsber. K. P. Akad. Wiss, Berlin*, 1864, pp. 59—70.

2. Ueber einige cephalopoden aus dem Muschelkalk der alpen und über verwandte Arten. [Ammonites from Himalayas]: *Abhandl. K. Akad. Wiss. Berlin*, 1866, pp. 105—150; *Monatsber. K. P. Akad. Wiss., Berlin*, 1865, pp. 660—672.

3. [African ammonites allied to Cutch species]: *Monatsber. K. P. Akad. Wiss. Berlin*, 1877, pp. 97—103.

Bhartpur.

1. Sketch of the geology of the Bhartpur District: *Glean. Sci.*, II, 143—147, (1830).

Bigge, H.

1. Notice of the discovery of coal and petroleum on the Namrup River: *Jour. As. Soc. Beng.*, VI, 243, (1837).

2. Despatch from Lieut. H. Bigge, Assistant Agent, detached to the Naga Hills, to Captain Jenkins, Agent, Governor General, North East Frontier. [Iron ore, hot springs, and limestones, near Golaghat]: *Jour. As. Soc. Beng.*, X, 129—136, (1841).

Bird, J.

1. A statistical and geological memoir of the country from Punah to Kittor, S. of the Krishna River: *Jour. Roy As. Soc.*, II, 65—80, (1835); *Mad. Jour. Lit. Sci.*, VI, 375—387, (1837).

Birdwood, G. C. M.
1. The Industrial arts of India. 8°. London, 1880.

Blackburn, C. H.
1. Experiments on the coal of Pind Dadan Khan : *Rec. Geol. Surv. Ind.,* XV, 63, (1882).

Blackwell, J. H.
1. Report of the Examination of the Mineral Districts of the Nerbudda valley : *Sel. Rec. Bo. Gov.,* XLIV, new series, (1857).

Blainville, H. D. de, *see* DE BLAINVILLE, H. D.

Blane, G. R.
1. Memoir on Sirmór : *Trans. Roy. As. Soc.,* I, 56—64, (1824) [mines, mostly in Jaonsar, on p. 61].

Blanford, H. F.
1. Notice of the occurrence of Crystalline Limestone in the district of Coimbatore : *Mad. Jour. Lit. Sci.,* XIX, new series, III, 60—65, (1857).

2. On the geological age of the sandstones containing Fossil wood, at Trivicary, near Pondichery : *Mad. Jour. Lit. Sci.,* XX, new series, IV, 47—53, (1858).

3. On the geological structure of the Nilghiri Hills, (Madras) : *Mem. Geol. Surv. Ind.,* I, pt. ii, 211—248, (1858).

4. Description of a native copper mine and smelting works in the Mahanaddi valley, Sikkim Himalaya : PERCY, *Metallurgy,* I, 388—392, (1861).

5. On Dr. Gerard's collection of fossils from the Spiti valley, in the Asiatic Society's Museum : *Jour. As. Soc. Beng.,* XXXII, 124—138, (1863).

6. On the Cretaceous and other rocks of the South Arcot and Trichinopoly Districts, Madras : *Mem. Geol. Surv. Ind.,* IV, pt. i, 1—217, (1863).

7. Remarks on rude stone Monuments in Chutia Nagpur : *Proc. As. Soc. Beng.,* 1863, pp. 130—131.

8. On a fossil amphibian (Labyrinthodon ?) from the Pachmarhi Hills, (Central India) : *Jour. As. Soc. Beng.,* XXXIII, 336—338, 442—444, 461—462, (1864).

9. Note on a tank section at Sealdah, Calcutta : *Jour. As. Soc. Beng.,* XXXIII, 154—158, (1864).

10. The fossil cephalopoda of the Cretaceous rocks of Southern India ; Belemnitidæ-Nautilidæ : *Pal. Indica,* 1st series, I, 1—40, (1865).

11. On the age and correlations of the Plant-bearing series of India and the former existence of an Indo-Oceanic continent : *Quart. Jour. Geol. Soc.,* XXXI, 519—542, (1875).

12. [On a neolithic celt from S. India] : *Proc. As. Soc. Beng.,* 1868, p. 59.

13. [Remarks on Himalayan glaciation] : *Proc. As. Soc. Beng.,* 1877, p. 3.

14. Rudiments of Physical Geography for the use of Indian Schools, and a glossary of the technical terms employed. 8°. *9th ed.* London, 1881.

Blanford, H. F., and W. T., and Theobald, W.
1. On the Geological Structure and relations of the Talcheer coalfield in the district of Cuttack : *Mem. Geol. Surv. Ind.,* I, pt. i, 33—89, (1856).

Blanford, H. F., *and* Salter, J. W.

1. Palæontology of Niti in the Northern Himalaya: being descriptions and figures of the palæozoic and secondary fossils collected by Col. Richard Strachey. 8°. Calcutta, 1865.

Blanford, H. F., *and* Stoliczka, F.

1. Catalogue of the specimens of Meteoric Stones and Meteoric Irons in the Museum of the Asiatic Society of Bengal, Calcutta, corrected up to January 1866: *Jour. As. Soc. Beng.*, XXXV, pt. ii, 43—45, (1866).

Blanford, W. T.

1. On the Geological structure and physical features of the districts of Bancoorah, Midnapore, and Orissa, Bengal. Note on the laterite of Orissa: *Mem. Geol. Surv. Ind.*, I, pt. iii, 249—279, (1859).

2. Report on the Beerbhoom Iron works. Flscp. Calcutta, 1860.

3. On the rocks of the Damuda group, and their associates in Eastern and Central India, as illustrated by the re-examination of the Raniganj field: *Jour. As. Soc. Beng.*, XXIX, 352—358, (1860).

4. Account of visit to Puppa doung, an extinct volcano in Upper Burma: *Jour. As. Soc. Beng.*, XXXI, 215—226, (1862); *Brit. Ass. Rep.*, 1860, pt. 2, pp. 69—70; Burma, pp. 341—350, (1882).

5. On the Geological structure and relations of the Raniganj coal-field, Bengal: *Mem. Geol. Surv. Ind.*, III, 1—195, (1863).

6. Report on the Pench river coal-field, in Chindwarrah district, Central Provinces: *Gazette of India Supplement*, 1866, pp. 367—377. See No. 52.

7. On worked agates of the early stone age from Central India: *Proc. As. Soc. Beng.*, 1866, pp. 230—234.

8. [On stone implements from Central India]: *Proc. As. Soc. Beng.*, 1867, pp. 136—138.

9. [On the superior antiquity of Indian stone weapons]: *Proc. As. Soc. Beng.*, 1867, pp. 144—145.

10. Note on the Geology of the neighbourhood of Lynyan and Runneekote, north-west of Kotree in Sind: *Mem. Geol. Surv. Ind.*, VI, 1—15, (1867).

11. On the Geology of a portion of Cutch: *Mem. Geol. Surv. Ind.*, VI, 17—38, (1867).

12. On the traps and Inter-trappean beds of Western and Central India: *Mem. Geol. Surv. Ind.*, VI, 137—162, (1867).

13. On the Geology of the Taptee and Lower Nerbudda valleys, and some adjoining districts: *Mem. Geol. Sur. Ind.*, VI, 163—394, (1869).

14. On the coal-seams of the Táwa valley: *Rec. Geol. Surv. Ind.*, I, 8—11, (1868); *Sel. Rec. Gov. Ind.*, LXIV, 31—35, (1868).

15. On the coal-seams of the neighbourhood of Chanda: *Rec. Geol. Surv. Ind.*, I, 23—26, (1868).

16. Coal near Nagpur: *Rec. Geol. Surv. Ind.*, I, 26, (1868).

17. Notes on route from Poona to Nagpur, *vid* Ahmednuggur, Jalna, Loonar, Yeotmahal, Mangali, and Hingunghát: *Rec. Geol. Surv. Ind.*, I, 60—65, (1868).

Blanford, W. T.,—cont.

18. Note on the lead vein near Chicholi, Raipur district: *Rec. Geol. Surv. Ind.*, III, 44—45, (1870); *Ind. Economist*, I, 363, (1870).

19. Report on the coal at Korba in Bilaspur district: *Rec. Geol. Surv. Ind.*, III, 54—57, (1870); *Ind. Economist*, II, 43—44, (1870).

20. On the occurrence of coal east of Chhatisgarh in the country between Bilaspur and Ranchi: *Rec. Geol. Surv. Ind.*, III, 71—72, (1870).

21. On faults in strata: *Geol. Mag.*, 1st decade, VII, 115—118, (1870).

22. Account of visit to the Eastern and Northern Frontiers of Independent Sikkim, with notes on the Zoology of the Alpine and Sub-Alpine regions: *Jour. As. Soc. Beng.*, XL, pt. ii, 367—420, (1871), XLI, pt. ii 30—71, (1872); *Proc. As. Soc. Beng.*, 1871, pp. 167—170, 226 —228.

23. Note on the plant-bearing sandstones of the Godavari valley, on the Southern extension of rocks belonging to the Kámthi group to the neighbourhood of Ellore and Rájámandri, and on the possible occurrence of coal in the same direction: *Rec. Geol. Surv. Ind.*, IV, 49—52, (1871).

24. Report on the progress and results of borings for coal in the Godavari Valley near Dumagudem and Bhadráchalam: *Rec. Geol. Surv. Ind.*, IV, 59-66, (1871); *Ind. Economist*, III, 45, (1871); *Gazette of India Supplement*, 1871, pp. 1117—1123.

25. Additional note on the plant-bearing sandstones of the Godavari valley: *Rec. Geol. Surv. Ind.*, IV, 82, (1871).

26. Description of sandstone in the neighbourhood of the first barrier on the Godávari, and in the country between the Godávari and Ellore: *Rec. Geol. Surv. Ind.*, IV, 107—115, (1871); V, 23—28 (1872).

27. Description of geology of Nagpur and its neighbourhood: *Mem. Geol. Surv. Ind.*, IX, 295—330, (1872).

28. Note on the geological formations seen along the coasts of Balúchistán and Persia from Karachi to the head of the Persian Gulf, and on some of the Gulf Islands: *Rec. Geol. Surv. Ind.*, V, 41—45, (1872).

29. Sketch of the geology of Orissa: *Rec. Geol. Surv. Ind.*, V, 56-65, (1872).

30. Note on Maskat and Massandim on the east coast of Arabia: *Rec. Geol. Surv. Ind.*, V, 75—77, (1872).

31. Sketch of the geology of the Bombay Presidency: *Rec. Geol. Surv. Ind.*, V, 82—102, (1872).

32. On the Nature and probable origin of the superficial Deposits in the valleys and Deserts of Central Persia: *Quart. Jour. Geol. Soc.*, XXIX, 493-503, (1873).

33. On some evidence of glacial action in tropical India in palæozoic (or the oldest mezozoic) times: *Brit. Ass. Rep.*, XLIII, pt. ii, 76, (1873).

34. On the physical geography of the deserts of Persia and Central Asia: *Brit. Ass. Rep.*, XLIII, pt. ii, 162-163, (1873).

35. On Flint cores and flakes from Sakhar and Rohri in Sind: *Proc. As. Soc. Beng.*, 1875, pp. 134—136.

36. Report on water-bearing strata of the Surat district: *Rec. Geol. Surv. Ind.*, VIII, 49—55, (1875).

Blanford, W. T.,—cont.

37. The Zoology and geology of Eastern Persia. 8°. London, 1876, Vol. II. [*See* GOLDSMID, F. J.]

38. On the Physical Geography of the great Indian Desert, with special reference to the former existence of the sea in the Indus Valley; and on the origin and mode of Formation of the Sand Hills: *Jour. As. Soc. Beng.*, XLV, pt. ii, 86-103, (1876); *Geol. Mag.*, 2nd decade, III, 507—511, (1876).

39. On the geology of Sind: *Rec. Geol. Surv. Ind.*, IX, 8-22, (1876).

40. Note on the geological age of certain groups comprised in the Gondwána series of India, and on the evidence they afford of distinct Zoological and Botanical Terrestrial Regions in ancient epochs : *Rec. Geol. Surv. Ind.*, IX, 79-85, (1876).

41. Dr. Feistmantel's paper on the Gondwána series : *Geol. Mag.*, 2nd decade, IV, 189, (1877).

42. Geological notes on the great Indian Desert between Sind and Rájputána : *Rec. Geol. Surv. Ind.*, X, 10-21, (1877).

43. The Palæontological relations of the Gondwana system : *Rec. Geol. Surv. Ind.*, XI, 109—150, (1878).

44. On the geology of Sind. (2nd notice) : *Rec. Geol. Surv. Ind.*, XI, 161—173, (1878).

45. [Geology of] the Konkan : BOMBAY, pp. 23-27.

46. [Geology of the] North-Western Deccan : BOMBAY, pp. 50-52.

47. Account of the geology of Sind, with an exhibition of a geological map : *Proc. As. Soc. Beng.*, 1878, pp. 3-8.

48. The geology of Western Sind : *Mem. Geol. Surv. Ind.*, XVII, 1—196, (1879).

49. Exhibition of a specimen of Hippuritic Limestone from Afghanistan : *Proc. As. Soc. Beng.*, 1879, p. 202.

50. Report on the proceedings and results of the Geological Congress at Bologna: *Rec. Geol. Surv. Ind.*, XV, 64—76, (1882).

51. Report on the Pench valley coalfield : *Rec. Geol. Surv. Ind.*, XV, 121—137, (1882). *See* No. 6.

52. Note of Mach in the Bolan Pass, and Sharág or Sharig on the Hurnai route between Quetta and Sibi : *Rec. Geol. Surv. Ind.*, XV, 149-153.

53. Geological notes on the hills in the neighbourhood of the Sind and Punjab frontier between Quetta and Dera Ghazi Khan : *Mem. Geol. Surv. Ind.*, XX, 105—240, (1883).

54. Presidential address: *Proc. As. Soc. Beng.*, 1879, pp. 33-62.

55. Presidential address to the geological section of the British Association, [Homotaxis as illustrated from Indian formations] : *Rep. Brit. Ass.*, 1884, pp. 691—711 ; *Rec. Geol. Surv. Ind.*, XVIII, 32—51, (1885).

56. Report on the International geological Congress of Berlin : *Rec. Geol. Surv. Ind.*, XIX, 13-22, (1886).

57. On additional evidence of the occurrence of glacial conditions in the Palæozoic era, and on the geological age of the Beds containing Plants of Mesozoic type in India and Australia : *Quart. Jour. Geol. Soc.*, 249—263, (1886).

Blanford, W. T.,—cont.

58. On a Smoothed and Striated Boulder from the Punjab Salt Range : *Geol. Mag.*, 3rd decade, III, 494-495, 574, (1886).

59. Note on a character of the Talchir boulder beds : *Rec. Geol. Surv. Ind.*, XX, 49, (1887).

Blanford, W. T. *and* Medlicott, H. B.

1. A manual of the geology of India, chiefly compiled from the observations of the Geological Survey : Pts. i and ii. 8°. Calcutta, 1879.

Blanford, W. T. *and* Stoliczka, F.

1. Scientific results of the second Yarkand Mission, based upon the notes and collections of the late Ferdinand Stoliczka. Geology. 4°. Calcutta, 1878. YARKAND, No. 1.

Blanford, W. T. *and* H. F., *and* Theobald, W.

1. On the geological structure and relations of the Talcheer coalfield in the District of Cuttack : *Mem. Geol. Surv. Ind.*, I, pt. i, 33—89, (1856).

Blochman, H.

1. Remarks on the Sundarban : *Proc. As. Soc. Beng.*, 1868, pp. 266-273.

2. Note on the fall of a Meteorite at Jullunder, in April, A.D. 1621 : *Proc. As. Soc. Beng.*, 1869, pp. 167—169.

Bloomfield, A.

1. Letter on pieces of copper and silver from Gungeria : *Proc. As. Soc. Beng.*, 1870, pp. 131—134.

Blundell, E. A.

1. An account of some of the petty states lying north of the Tenasserim Provinces, drawn up from the Journals and Reports of Dr. D. Richardson : *Jour. As. Soc. Beng.*, V, 601—625, 688—707, (1836).

Boileau, J. T.

1. Observations on the Sandstones of the quarries near Agra and on the results of experiments made thereon : *Glean. Sci.*, II, 158—160, (1830).

Bolan pass.

Letter on the Bolan pass : *Jour. Roy. Geog. Soc.*, XII, 109.

Bombay.

1. Articles on the Geology of Portions of the Bombay Presidency, written for the Bombay Gazetteer. 8vo. Bombay, 1878. Contains, W. T. BLANFORD, Nos. 46, 47 ; R. B. FOOTE, Nos. 11, 12, 13 ; A. LEITH, No. 2 ; T. THEOBALD, No. 27.

Boring.

1. Report on the Experimental Boring for Fresh Water in Fort William : *Glean. Sci.*, III, 124-125, (1831).

2. Report of the Committee, appointed on the 27th March, 1833, to consider on the expediency of recommending to the Government the continuance of the Boring Experiment : *Jour. As. Soc. Beng.*, II, 369-374, (1833)

3. Account of an attempt to form an Artesian·Well at Tuticorin. From official papers : *Mad. Jour. Lit. Sci.*, XV, 167—172, (1848).

Boring,—cont.

4. Boring for water at the Red Hills, [Madras, Section in—] : *Ind. Jour. Arts Sci.*, 2nd series, I, 162, (1857).

Bose, P. N.

1. Undescribed Fossil Carnivora from the Sivalik Hills in the collection of the British Museum : *Quart. Jour. Geol. Soc.*, XXXVI, 119—136, (1880).

2. Undescribed Fossil Carnivora from the Sivaliks in the collection of the British Museum : *Rec. Geol. Surv. Ind.*, XIV, 263—267, (1881).

3. Note on some Earthen Pots found in the alluvium at Mahesvara (Mahesur) : *Jour. As. Soc. Beng.*, LI, pt. i., 226-229, (1882).

4. Geology of the Lower Narbaddá valley between Nimáwar and Kawant : *Mem. Geol. Surv. Ind.*, XXI, 1-72, (1884).

5. Note on Lignite near Raipur, Central Provinces : *Rec. Geol. Surv. Ind.*, XVII, 130-131, (1884).

6. The Iron Industry of the Western Portion of the District of Raipur : *Rec. Geol. Surv. Ind.*, XX, 167—170, (1887).

Bournon, Count de.

1. Description of the Corundum Stone and its varieties, commonly known by the names of Oriental Ruby, Sapphire, &c ; with observations on some other mineral substances : *Phil. Trans.*, 1802, pp. 233—326.

2. A descriptive catalogue of diamonds in the cabinet of Sir Abraham Hume, Bart. 4°. London, 1815.

3. Observations sur quelques-uns des Minéraux, soit de l'Ile de Ceylan soit de la côte de Coromandel, rapportés par M. Lerchenault de Latour. 4°. Paris, 1823.

3. Description of certain gangues of spinelle brought from the Island of Ceylon : *Phil. Mag.*, LXIII, 30—36, (1824).

Boyle, A.

1. [Account of an eruption off False Island, Ramree] : *Jour. As. Soc. Beng.*, XIII, Proc., p. xxxv, (1844).

Boyle, J. A.

1. Nellore District Manual. 8°. Madras, 1873, [Geology and Soils, compiled from notes furnished by C. Æ. Oldham, Soils by Mr. Charles Rundall ; pp. 40—59].

Bradley, W. H.

1. Some account of the topography and climate of Chiculda, situated on the table-land of the Gawil Range : *Trans. Bo. Geog. Soc.*, VII, 167—185, (1845).

2. Notice of a native carbonate of soda found in the territory of the Nizam, India : *Pharmaceut. Jour.*, XII, 515—516, (1853).

Brandis, D.

1. Suggestions regarding the management of the leased forests of Busahir in the Sutlej valley of the Punjab. Flscp. Simla, 1881.

2. Report of—upon the Deodar Forests of Busahir. Flscp. Calcutta, 1865

Branfill, R. B.

1. Notes on the Tinevelly District. Dehra Doon, 1869.

2. Notes on the Physiography of Southern India: *Proc. Roy. Geog. Soc.,* new series, VII, 719—735, (1885): *Brit. Ass. Rep.,* 1885, pp. 1124—1126.

Breton, P.

1. Description of the animals and reptiles met with in the Districts of Ramgur, Surgoojah, and Sumbhulpore, and of the principal mineral productions of these provinces: *Trans. Med. Phys. Soc. Calcutta,* II, 246—274, (1826).

2. Account of the diamond workings and diamonds of Sumbhulpore: *Edin. Jour. Sci.,* VII, 134—140, (1827); *Froriep. Notizen.,* XVIII, 145—150, (1827).

Briggs, D.

1. Memorandum on boulders in India: *Mad. Jour. Lit. Sci.,* XIII, 188—189, (1844).

2. Report on the operations connected with the Hindustan and Thibet Road: *Sel. Rec. Gov. Ind.,* XVI, 11—13, (1856).

Brooke, H. J.

1. On Poonahlite, a new species of mineral; on the identity of Zeagonite and Philliprite, &c., and other Mineralogical notices: *Phil. Mag.,* X, 109—112, (1831).

Brooke, J. C.

1. Note on the Zinc Mines of Jawar: *Jour. As. Soc. Beng.,* XIX, 212—215, (1850).

2. The Mines of Khetree in Rajputana: *Jour. As. Soc. Beng.,* XXXIII, 519—629, (1864).

Broome, A. *and* Cunningham, Alex.

1. Abstract Journal of the routes of Lieutenants A. Broome and A. Cunningham to the sources of the Punjab rivers: *Jour. As. Soc. Beng.,* X, 1—6, (1841).

Broughton, F.

1. On the transition of Trap into Laterite: *Jour. Bo. As. Soc.,* V, 639—642, (1857).

Broun, J. A.

1. On Magnetic Rocks in South India: *Brit. Ass. Rep.,* 1860, pt. ii, pp. 24—27.

2. On the velocity of earthquake shocks in the Laterite of India: *Brit. Ass. Rep.,* 1860, pt. ii, pp. 74-75.

Brown, R.

1. Annual Report on the Manipur Political Agency: *Sel. Rec. Gov. Ind.,* LXXVIII, (1870).

Brownlow, C.

1. Note on the occasional existence of fresh water on the surface of the ocean: *Jour. As. Soc. Beng.,* V, 239, (1836).

Brownlow, H. A.

1. Report on the means and advisability of manufacturing artificial Hydraulic cements in the North-Western Provinces of India: *Prof. Pap. Ind. Eng.*, 1st series, VII, 400—412, (1870).

Brownlow, H. A. *and* Desjoux, P.

1. Notes on the proposed manufacture of Hydraulic cements in India: *Prof. Pap. Ind. Eng.*, 2nd series, I, 604—621, (1872).

Brownlow, H. A., Higham, T. *and* Nielly, A.

1. Extracts from reports and letters on kunkur limes and cements on the Bari Doab canal: *Prof. Pap. Ind. Eng.*, 2nd series, VI, 127—168, (1877).

Bryce, J. A.

1. Burma: the Country and the people; *Proc. Roy. Geog. Soc.*, new series, VIII, 481—499, (1886). [Contains some notes on the hills west of the Kubo valley.]

Buchanan, Francis [*afterwards* Hamilton].

1. A Journey from Madras through the countries of Mysore, Canara, and Malabar. 3 vols. 4°. London, 1807.

2. Description of the Diamond mine at Panna: *Edin. New Phil. Jour.*, I, 49—54, (1819).

3. Account of the Mine or Quarry of corundum in Singraula: *Edin. New Phil. Jour.*, II, 305—307, (1820).

4. On the Minerals of the Rajmehal cluster of hills: *Glean. Sci.*, III, 1—8, 33—39, (1831).

Buckland, W.

1. Geological account of a series of Animal and Vegetable remains and of Rocks, collected by J. Crawfurd, Esq., on a voyage up the Irawadi to Ava in 1826-27: *Geol. Trans.*, 2nd series, II, 377—392, (1829); *Proc. Geol. Soc.*, 1834, pp. 71—73; abstracted in *Glean. Sci.*, I, 184—186, (1829); Crawfurd, *Embassy to Ava*, 1st ed., pp. 78—88, 2nd ed., II, pp. 143—162, (1834).

2. Supplementary remarks on the supposed power of the waters of the Irrawadi to convert wood into stone: *Geol. Trans.*, 2nd series, II, 403-404, (1829); *Edin. New Phil. Jour.*, VI, 67—70, (1828).

Buist, G.

1. Note on a set of specimens from Aden: *Jour. Bo. As. Soc.*, I, 344-345, (1843).

2. Note on a series of Persian Gulf specimens: *Jour. Bo. As. Soc.*, I, 345-346, (1843).

3. Annual address to the Bombay Geographical Society: *Trans. Bo. Geog. Soc.*, IX, pp. li—cix, (1850).

4. Notices of the most remarkable Meteors in India, of the fall of which accounts have been published: *Trans. Bo. Geog. Soc.*, IX, 197—231, (1850).

Buist, G.,—cont.

5. On the general vibration or descent and upheaval which seems, at a recent geological period, to have occurred all over the Northern Hemisphere: *Jour. As. Soc. Beng.*, XIX, 300—309, (1850) ; *Edin. New Phil. Jour.*, L, 322—329, (1851); *Brit. Ass. Rep.*, 1851, pt. ii,. pp. 55—58.

6. Floods in India in 1849: *Jour. As. Soc. Beng.*, XX, 186—192, (1851) ; *Edin. New Phil. Jour.*, L, 52—58, (1851).

7. Index to books and papers on the Physical geography, antiquities, and statistics of India. 8°. Bombay, (1852).

8. The Volcanoes of India: *Trans. Bo. Geog. Soc.*, X, 139—167, (1852); *Edin. New Phil. Jour.*, LII, 339—352, (1852), LIII, 32—38, (1853); WESTERN INDIA, 169—206, (1857).

9. The Geology of Bombay: *Trans. Bo. Geog. Soc.*, X, 167—239, (1852).

10. On the Physical geography of Hindustan: *Edin. New Phil. Jour.*, 1st series, LVI, 328—353, (1854).

11. Notes on a journey through parts of Kattywar and Guzerat: *Trans. Bo. Geog. Soc.*, XIII, 11—107, (1855).

12. On the principal depressions on the surface of the globe: *Edin. New Phil. Jour.*, 2nd series, I, 253—264, (1856).

13. On the Physical geography of India: *Trans. Bo. Geog. Soc.*, XIV, pp. vi—xxi, (1859).

14. On the Physical geography of the Red Sea: *Trans. Bo. Geog. Soc.*, XIV, 1—40, (1856).

15. Remarks on the Laterite of Cochin and Quilon : *Trans. Bo. Geog. Soc.*, XV, pp. xxii—xxiv, (1860).

Bunbury, *Sir* C. J. F.

1. On a remarkable specimen of Neuropteris, with remarks on the genus: *Quart. Jour. Geol. Soc.*, XIV, 243—249, (1858).

2. Notes on a collection of Fossil plants from Nagpur, India: *Quart. Jour. Geol. Soc.*, XVII, 325—354, (1861).

Burma.

1. Papers on the Geology and minerals of British Burmah, reprinted by order of C. E. Bernard, C.S.I., Chief Commissioner. 8°. Calcutta, 1882 ; Contains W. T. BLANFORD, No. 4; G. D'AMATO, No. 1; M. FRYAR, Nos. 4, 5, 6, 7, 8 ; F. R. MALLET, Nos. 14, 15, 16, 21, 26 ; T. OLDHAM, Nos. 1, 8, 9, 10; E. O'RILEY, No. 5; THEOBALD, Nos. 7, 10, 11, 12, 13, 15, 16, 18, 19; J. B. TREMENHEERE, Nos. 2, 3, 8 ; WHITE, No. 1.

Burnes, *Sir* A.

1. A memoir and supplementary memoir of a map of the Eastern Branch of the Indus. [Lithographed.] Flscp. Bombay, 1828.

2. On the geology of the Bank of the Indus, the Indian Caucasus, and the plains of Tartary to the shores of the Caspian: *Geol. Trans.*, 2nd series, III, 491—494, (1835); *Quart. Jour. Geol. Soc.*, II, 8—10, (1831); *Proc. Geol. Soc.*, II, 8—10, (1838).

3. Comparison of the Indus and Ganges rivers: *Jour. As. Soc. Beng.*, I, 21—23, (1832).

Burnes, *Sir* A.,—cont.

  4. Notice of an earthquake at Lahore, 22nd January, 1832 : *Jour. As. Soc. Beng.*, I, 34, (1832).

  5. Some account of the Salt mines of the Punjab : *Jour. As. Soc. Beng.*, I, 145—148, (1832).

  6. Peshawar coal : *Jour. As. Soc. Beng.*, II, 267, (1833).

  7. Description of the salt works at Panchpadder, Méwar : *Jour. As. Soc. Beng.*, II, 365—366, (1833).

  8. Substance of a geographical memoir on the Indus : *Jour. Roy. Geog. Soc.*, III, 287—289, (1833).

  9. Papers descriptive of the countries on the north-west Frontier of India :—
The Thurr or Desert, Joodpoor, and Jaysulmeer : *Jour. Roy. Geog. Soc.*, IV, 88—128, (1834).

 10. Memoir on the eastern branch of the river Indus, giving an account of the alterations produced on it by an earthquake ; also a Theory of the formation of the Runn and some conjectures on the Route of Alexander the Great : *Trans. Roy. As. Soc.*, III, 550—588, (1834).

 11. Travels in Bokara, containing the narrative of a voyage on the Indus, from the sea to Lahore, with presents from the king of Great Britain : and an account of a journey from India to Cabool, Tartary, and Persia. (2nd edition.) 3 vols., 12°. London, 1835.

 12. Presentation and a brief account of a series of the Geology and Fossil Conchology of the Cheri range in Cutch : *Jour. As. Soc. Beng.*, VI, 159, (1837).

 13. On the Reg-Ruwan or moving sand : a singular phenomenon of sound, near Cabul : *Jour. As. Soc. Beng.*, VII, 324—325, (1838).

Burnes, Alexander, *and* Gerard, J. G.

  1. A sketch of the Route and progress of Lieut. A. Burnes and Dr. Gerard : *Jour. As. Soc. Beng.*, I, 139—145, (1832) ; II, 1—22, 143—149, (1833).

Burr, F.

  1. Remarks on the auriferous deposits of India considered with special reference to their economic value : *Mad. Jour. Lit. Sci.*, XII, 30—37, (1840) ; *Proc. Geol. Soc. Lond.*, III, 355—356, (1842).

  2. Sketch of the Geology of Aden on the Coast of Arabia : *Geol. Trans.*, 2nd series, VI, 499—502, (1842).

Burt, T. S.

  1. Description of the Mode of Extracting salt from the damp sand-beds of the River Jumna, as practised by the inhabitants of Bundelkhand : *Jour. As. Soc. Beng.*, 33—36, (1834).

Burton, *Capt.* R. F.

  1. The Nizam Diamond. The Diamond in India : *Quart. Jour. Sci.*, VI, new series, 351—360, (1876).

Busk, G.

  1. Has the Asiatic elephant been found in a fossil state ? [by A. L. Adams], with additional remarks by G. Busk : *Quart. Jour. Geol. Soc.*, XXIV, 496—499, (1868) ; *Phil. Mag.* XXXVII, 152, (1869).

Butler, J.
  · 1. Earthquakes in Assam : *Jour. As. Soc. Beng.*, XVIII, 172—175, (1849).

Bysack, Gaur Dass.
  1. On the Gopalpore aërolite : *Proc. As. Soc. Beng.*, 1865, p. 94.

C

Calcutta.
  1. Salt Water lakes in the vicinity of Calcutta, with suggestions for filling them up by warping : *Glean. Sci.*, II, 201—207, (1830).

Calder, J.
  1. General observations on the Geology of India : *As. Res.*, XVIII, pt. i, 1—22, (1833); *Edin. New Phil. Jour.* VI, 152—188, (1829) ; *Glean. Sci.*, I, 211—213, (1829).
  2. Note. on certain specimens of animal remains from Ava : *Glean. Sci.*, I, 167—170, (1831).

Calthrop, C. W.
  1. Report on the hot springs of Kulu : *Kulu Gazetteer.* 2 vols., 8°. Punjab Government, 1883—84 ; Vol. II, Appendix, pp. 150—153.

Calvert, John.
  1. Notes on the mineral resources of India. 8°. Calcutta, (1870 ?)
  2. Kulu : its Beauties, Antiquities, and Silver mines. 8°. Calcutta, (1871).

Campbell, D. A.
  1. On the notice of Alum or Salájit of Nepal : *Jour. As. Soc. Beng.*, II, 482—484, (1833).
  2. Account of the earthquake at Kathmandu, 26th Aug. 1833 : *Jour. As. Soc. Beng.*, II, 438—439, 564—567, and 636—639, (1833).
  3. On the Nepalese method of refining gold : *Jour. As. Soc. Beng.*, III, 622—627, (1834).
  4. Journal of a trip to Sikkim, in December 1848, with a sketch map : *Jour. As. Soc. Beng.*, XVIII, 482—541, (1849)
  5. A journey through Sikkim to the frontiers of Thibet : *Jour. As. Soc. Beng.*, XXI, 407—428, 477—501, and 563—575, (1852).
  6. Notes on Eastern Thibet : *Jour. As. Soc. Beng.*, XXIV, 215—240, (1855).

Campbell, A. *and* Piddington, H.
  1. [Correspondence respecting the discovery of copper ore at Pushak, near Darjeeling] : *Jour. As. Soc. Beng.*, XXIII, 206—210, (1854) ; XXIV, 251, 707—708, (1855).

Campbell, C.
  1. Account of the slate quarries in the Goorgaon District : *Prof. Pap. Ind. Eng.*, IV, 257—260, (1867).

Campbell, J.
  1. Remarks on nomenclature of Indian minerals contained in a paper on "Chemical Tests," by Lieutenant Braddock : *Mad. Jour. Lit. Sci.*, X, 270—271, (1839).

Campbell, J.,—cont.

2. On the advancement of Geological Science in India : *Mad. Jour. Lit. Sci.*, XI, 78—86, (1840).

3. On an error in Dr. Thomson's Mineralogy : *Mad. Jour. Lit. Sci.*, XI, 310—313, (1840).

4. Report on the soda soils of the Barramahal : *Jour. As. Soc. Beng.*, X, 159—162, (1841) ; *Trans. Bo. Geog. Soc.*, VI, 163—167, (1840).

5. Report on the Kaolin earth of Mysore : *Jour. As. Soc. Beng.*, X, 163—164, (1841) ; *Trans. Bo. Geog. Soc.*, VI, 166—168, (1840).

6. Report on the Manufacture of Steel in Southern India : *Jour. As. Soc. Beng.*, XI, 217, (1842).

7. On the Red Marl formation of Mysore : *Cal. Jour. Nat. Hist.*, II, 32—42, (1842).

8. On the granite formation of the Salem and Barramahal Districts: *Cal. Jour. Nat. Hist.*, II, 153—185, (1842).

9. Extract from letter [relating to Pondicherry fossils] : *Cal. Jour. Nat. Hist.*, II, 276—279, (1842).

10. Extract from report addressed to the Coal and Mineral Committee : *Cal. Jour. Nat. Hist.*, II, 280—283, (1842).

11. On the Schistose formations of the Table-lands of South India, with the characters of Hornblendic rocks : *Cal. Jour. Nat. Hist.*, II, 301—322, (1842).

12. Suggestions regarding the probable origin of some kinds of kunkur and the influence of deliquescent salts on vegetation : *Cal. Jour. Nat. Hist.*, III, 25—28, (1843).

13. On the Manufacture of bar iron in Southern India : *Cal. Jour. Nat. Hist.*, III, 386—400, (1843) ; V, 103—115, (1845) ; VI, 34—43, (1846); *Appendix, Rep. Gov. Museum, Madras,* pp. 8—17, 19—28, (1856).

14. On the Mineralogy of Southern India : *Cal. Jour. Nat. Hist.*, VI, 199—200, (1846).

Campbell, J. F.

1. On Himalayan glaciation : *Jour. As. Soc. Beng.*, XLVI, pt. ii, pp. 1—10, (1877) ; *Proc. As. Soc. Beng.*, 1877, pp. 2-5.

Campbell, J. M.

1. Account of the Rajpipla Cornelians : *Bombay Gazetteer*, VI, 198—207, (1880).

Cantor, T.

1. Notice of a skull (fragment of a gigantic fossil Batrachian) : *Jour. As. Soc. Beng.*, VI, 538—541, (1837).

Carey, J. J.

1. On stone celts from Khangaon : *Supplement, C. P. Gazette,* 4th September 1869; *Proc. As. Soc. Beng.*, 1871, pp. 238—239.

Carey, V. J.

1. On stone spindle whorls : *Proc. As. Soc. Beng.*, 1866, p. 135.

22

Carless, T. G.

1. Memoir to accompany the survey of the delta of the Indus in 1837 : *Jour. Roy. Geog. Soc.*, VIII, 328—366, (1838) ; *Sel. Rec. Bo. Gov.*, XVII, 459—500, (1855).

2. Visit to the hot spring near Kurrachee : *Trans. Bo. Geog. Soc.*, Feb. 1839, pp. 13—18.

3. Report upon portions of the river Indus : surveyed in the years 1836—37 ; accompanied by a journal kept by that officer during that period : *Sel. Rec. Bo. Gov.*, XVII, 501—540, (1855).

Carnac, H. Rivett.

1. [Stone implements exhibited by—] : *Proc. As. Soc. Beng.*, 1882, pp. 6—8.

2. Note on some geological specimens received from Prof. Dr. Fischer : *Proc. As. Soc. Beng.*, 1883, p. 79.

3. On stone implements from the North-Western Provinces of India : *Jour. As. Soc. Beng.*, LII, pt. i, 221—230, (1883).

Carte, Alexander.

1. On a fossil elephant's tooth obtained from the excavation of the Doab canal in Upper India : *Jour. Geol. Soc. Dublin*, VIII, 66—68, (1857—60) ; *Nat. Hist. Rev.*, V, *Proc.*, pp. 84—86, (1858).

Carter, H. J.

1. Geological observations on the composition of the hills and alluvial soil from Hydrabad in Sindh, to the mouth of the river Indus : *Jour. Bo. As. Soc.*, II, 40—43, (1848).

2. Report on Copper ore and Lithographic Limestone from the S.-E. coast of Arabia : *Jour. Bo. As. Soc.*, II, 400, 1848.

3. On Foraminifera, their organization, and their existence in a fossilized state in Arabia, Sindh, Kutch and Khattyawar : *Jour. Bo. As. Soc.*, III, Pt. ii, 158—173, (1849).

4. Geological observations on the Igneous Rocks of Maskat and its neighbourhood, and on the Limestone Formation at their circumference : *Jour. Bo. As. Soc.*, III, pt. ii, 118—129, (1851).

5. Description of *Orbitolites malabarica*, illustrative of the spiral, and not concentric arrangement of the chambers in d'Orbigny's order *Cyclostègues* : *Jour. Bo. As. Soc.*, V, 142—145, (1857) ; *Ann. Mag. Nat. Hist.*, 2nd series, XI, 425—427, (1853).

6. Description of some of the larger forms of Fossilised *Foraminifera* in Sind, with observations on their internal structure : *Jour. Bo. As. Soc.*, V, 124—142, (1857) ; *Ann. Mag. Nat. Hist.*, 2nd series, XI, 161—177, 1853 ; with alterations and additions in WESTERN INDIA, [*infra* No. 14] 533—550, (1857).

7. Memoir on the geology of the South-East Coast of Arabia : *Jour. Bo. As. Soc.*, IV, 21—96, (1853) : WESTERN INDIA, [*infra* No. 14], 551—627, (1857).

8. Geology of the Island of Bombay : *Jour. Bo. As. Soc.*, IV, 161—216, (1853) ; WESTERN INDIA, (*infra* No. 14), 116—168, (1857).

9. Note on the pliocene deposits of the shores of the Arabian Sea : *Jour. Bo. As. Soc.*, IV, 445—448, (1853).

Carter, H. J.—cont.

10. On the true position of the canaliferous structure in the shell of Fossil *Alveolina*, d Orbigny; *Ann. Mag. Nat. Hist.*, XIV, 2nd series, 99—101, (1854).

11. Summary of the geology of India between the Ganges, the Indus, and Cape Comorin: *Jour. Bo. As. Soc.*, V, 179—336, (1857); WESTERN INDIA, (*infra* No. 14), pp. 628—776, (1857).

12. On contributions to the geology of Central and Western India: *Jour. Bo. As. Soc.*, V, 614—638, (1857). [Consists of several minor notices indexed under their various authors.]

13. Notice of D'Archiac and Haimé's Groupe Nummulitique de l'Inde: *Jour. Bo. As. Soc.*, V, 628—637, (1857).

14. Geological papers on Western India, including Cutch, Sind, and the south-east coast of Arabia; to which is appended a summary of the geology of India generally. 1 vol., 8°. With folio atlas. Bombay, 1857. *See* WESTERN INDIA.

15. Report on geological specimens from the Persian gulf, collected by C. G. Constable: *Jour. As. Soc. Beng.*, XXVIII, 41—48, (1859), XXIX, 359—365, (1860).

16. On the structure of the larger Foraminifera: *Ann. Mag. Nat. Hist.*, 3rd series, VIII, 246—251, (1861).

17. Further observations on the structure of Foraminifera and on the larger specialised forms of Sind, etc.: *Ann. Mag. Nat. Hist.*, 3rd series, VIII, 309—333, 366—382, 416—470, (1861); *Jour. Bo. As. Soc.*, VI, 31—96, (1861).

18. On contributions to the geology of Western India, including Sind and Baluchistan: *Jour. Bo. As. Soc.*, VI, 161—206, (1861). [Records various minor notices, indexed under their respective authors.]

19. Notes on the geology of the Islands around Bombay: *Jour. Bo. As. Soc.*, VI, 167—178, (1861).

20. Discovery of a portion of Pegmatite in a Basaltic dyke in the Island of Caringa: *Jour. Bo. As. Soc.*, VI, 178—180, (1861).

21. Section of the trap in the Western Ghats: *Jour. Bo. As. Soc.*, VI, 181—182, (1861).

22. Notes on the fossil bones from Narrayanpur, near Saugor, Central India: *Jour. Bo. As. Soc.*, VI, 204, (1861).

Cautley, *Sir* P. T.

1. Notice of the occurrence of Coal and Lignite in the Himalayas: *As. Res.*, XVI, 387—396, (1828).

2. On gypsum in the Himalaya: *Jour. As. Soc. Beng.*, I, 289—296, (1832).

3. Discovery of an ancient town near Behut, in the Doab: *Jour. As. Soc. Beng.*, III, 43—44, (1834).

4. Further account of the remains of an ancient town discovered at Behat, near Seháranpur: *Jour. As. Soc. Beng.*, III, 221—229, (1834).

5. [Letters regarding Sewalik fossils and buried town at Behat]: *Jour. As. Soc. Beng.*, III, 527—528, 592—593, (1834); IV, 585—587, (1835).

6. Note on the gold-washing of the Gúmti River: *Jour. As. Soc. Beng.*, IV, 279—281, (1835).

7. Note on the fossil Crocodile of the Seválik Hills: *As. Res.*, XIX, pt. i, 25—31, (1836).

Cautley, *Sir* **P. T.,**—cont.

8. The fossil gharial of the Siválik Hills: *As. Res.,* XIX, pt. i, 32—38, (1836).

9. Note on the teeth of the Mastodon à dents étroites of the Siwalik Hills: *Jour. As. Soc. Beng.,* V, 294—295, (1836).

10. Note on Mastodons of the Sewaliks: *Jour. As. Soc. Beng.,* V, 768—769, (1836).

11. On the remains of mammalia found in the Sewálik mountains, at the southern foot of the Himalayas between the Sutlej and the Ganges: *Proc. Geol. Soc.,* II, 395—397, (1838).

12. On the discovery of quadrumanous remains in the Sewálik Hills: *Proc. Geol. Soc.,* II, 544—545, (1838).

13. On the discovery of remains of a fossil monkey in the Sewalik Hills: *Proc. Geol. Soc.,* II, 568—569, (1838).

14. Note on a fossil Ruminant genus allied to Giraffidæ in the Siwalik Hills: *Jour. As. Soc. Beng.,* VII, 658—659, (1838); *Ann. Mag. Nat. Hist.,* ıst series, III, 167—169, (1839).

15. On the fossil remains of Camelidæ of the Sewaliks: *Jour. As. Soc. Beng.,* IX, 620—623, (1840).

16. On the structure of the Sewalik Hills and the organic remains found in them: *Geol. Trans.,* 2nd series, V, 267—278, (1840); *Mad. Jour. Lit. Sci.,* XII, 292—304, (1840).

17. Report on the influence of the Jumna Canals on the Jumna River: *Sel. Pub. Corr. N.-W. P.,* No. XLII, 137—142, (1849); *Sel. Rec. Gov N.-W. P.,* II, 57—60, (1855); *ibid.,* new series, IV, 490—496, (1868).

18. Report on the Ganges canal works; from their commencement until the opening of the Canal in 1854. 3 vols., with folio atlas. 8° and 4°. London, 1860.

Cautley, *Sir* **P. T.** *and* **Falconer, Dr. H.**

1. Synopsis of fossil genera and species from the upper deposits of the tertiary strata of the Siválik Hills, in the collection of the authors: *Jour. As. Soc. Beng.,* IV, 706—707, (1835).

2. Sivatherium giganteum, a new Fossil Ruminant genus from the valley of the Markanda, in the Siválik Branch of the Sub-Himalayan Mountains: *As. Res.,* XIX, pt. i, pp. 1—24, (1836): *Jour. As. Soc. Beng.,* V, 35—50, (1836); *Ann. Nat. Sci. Paris,* (*Zool.*), V, 348—370, (1836); *Bibl. Univ.,* IV, 392—397, (1836).

3. Note on the Fossil Hippopotamus of the Siválik Hills: *As. Res.,* XIX, pt. i, pp. 39—53, (1836).

4. Note on the Fossil Camel of the Siválik Hills: *As. Res.,* XIX, pt. i, pp. 115—134, (1836).

5. Note on the Felis Cristata, a new fossil Tiger, from the Sewálik Hills: *As. Res.,* XIX, pt. i, 135—192, (1836).

6. Note on the Ursus Sivalensis, a new fossil species from the Siwálik Hills: *As. Res.,* XIX, pt. i, pp. 193—201, (1836).

7. Notice on the Remains of a Fossil Monkey from the Tertiary Strata of the Sewálik Hills in the North of Hindoostan: *Geol. Trans.,* 2nd series, V, 499—504, (1840); *Proc. Geol. Soc.,* II, 568—569, (1838); *Mad. Jour. Lit. Sci.,* XII, 304—309, (1840).

Cautley, *Sir* **P. T.** *and* **Falconer, Dr. H.,**—cont.

8. On additional Fossil Species of the order Quadrumana from the Sewálik Hills : *Jour. As. Soc. Beng.*, VI, 354—360, (1837).

9. Communication on the Colossochelys atlas : *Proc. Zool. Soc.*, XII, 54—55, 84—88, (1844).

10. On some fossil remains of Anoplotherium and Giraffe from the Sewalik Hills in the north of India : *Proc. Geol. Soc.*, IV, 235—249, (1846) ; *Ann. Mag. Nat. Hist.*, 1st series, XIV, 501—502, (1844); XV, 55—59, (1845) ; *Cal. Jour. Nat. Hist.*, V, 577—589, (1845).

11. Fauna antiqua Sivalensis, being the fossil Zoology of the Siwalik Hills in the north of India. 8°, with folio atlas. London, 1846.

Center, W.

1. Note on Reh or alkali soils and saline well waters : *Rec. Geol. Surv. Ind.*, XIII, 253—273, (1880).

Chamier, H.

1. [Minute on the production of Bar iron in Southern India] : *Appendix, Rep. Gov. Mus. Madras*, 1856, pp. 29—34,

Chaper.

1. De la présence du diamant dans une pegmatite de l Indoustan : *Comptes Rendus*, XCVIII, 113—115, (1884).

2. Note sur une pegmatite diamantifère de l'Hindoustan : *Bull. Soc. Géol. France*, 3rd series, XIV, 330—345, (1886).

Cheduba.

1. Eruption of a mud volcano [on 31st Dec., 1881] : *Rec. Geol. Surv. Ind.*, XV, 141-142, (1882).

2. Fiery eruption from one of the mud volcanoes of Cheduba [on 23rd March, 1883] : *Rec. Geol. Surv. Ind.*, XVI, 204-205, (1883).

3. Notice of a fiery eruption from one of the mud volcanoes of Cheduba [on 3rd July, 1886] : *Rec. Geol. Surv. Ind.*, XIX, 268, (1886).

Chenevix, R.

1. Analysis of Corundum, and of some of the substances which accompany it ; with observations on the affinities which the earths have been supposed to have for each other, in the humid way : *Phil. Trans.*, 1802, 327—347 ; *Jour. de Phys.*, LV, 409—426, (1802).

Chevalier, E.

1. Voyage autour du monde exécuté pendant les années 1836 et 1837, sur corvette "*La Bonite,*" commandée par M. Vaillant. Géologie et Minéralogie. 8°. Paris, (1844). [Cretaceous Rocks of S. India.]

Christie, A. T.

1. Sketches of the Meteorology, Geology, Agriculture, Botany and Zoology of the southern Mahratta country : *Edin. New Phil. Jour.*, V, 292—304, VI, 98—101, (1828); VII, 49—65, (1829) ; *Mad. Jour. Lit. Sci.* IV, 185—193, 452—483, (1836); Geology in WESTERN INDIA, pp. 328—345, (1857).

2. On the Porcelain clay found at Mangalore : *Jour. As. Soc. Beng.*, X, 967—971, (1841).

Christison, R.

1. Chemical examination of the Petroleum of Rangoon: *Trans. Roy. Soc. Edin.*, XIII, 118—123, (1836).

Christopher, W.

1. Account of Adam's bridge and Ramiseram Temple; with a map of the said Temple: *Trans. Bo. Geog. Soc.*, VII, 130—138, (1846).

2. Report of an experimental voyage up the Indus and Sutledge: *Trans. Bo Geog. Soc.*, VIII, 144—173, (1849).

3. Ascent of the river Chenab: *Trans. Bo. Geog. Soc.*, VIII, 236—248, (1849).

Clark, G. T.

1. On the neighbourhood of Bombay and certain beds containing Fossil Frogs: *Quart. Jour. Geol. Soc.*, III, 221—224, (1847).

2. Remarks upon the Basalt Dykes of the mainland of India opposite to the Islands of Bombay and Salsette: *Quart. Jour. Geol. Soc.*, XXV, 163—168, (1869).

3. On volcanic foci of eruption in the Konkan: *Rec. Geol. Surv. Ind.*, XIII, 69—73, (1880).

Clark, Hyde.

1. Gold in India: *Jour. Soc. Arts*, XXIX, 243—256, (1881); XXX, 590, (1882).

Clark, J.

1. On the Lateritic formation: *Mad. Jour. Lit. Sci.*, VIII, 334—346, (1838).

2. Geology of Bangalore and of some other portions of Mysore: *Mad. Jour. Lit. Sci.*, IX, 89—121, (1839).

Clibborn, J.

1. Irrigation from Wells in the N.-W. Provinces and Oudh: *Prof. Pap. Ind. Eng.*, 3rd series, I, No. xiv, 103—132, (1883); II, No. lii, 105—109, (1884).

Clift, W.

1. On the Fossil remains of two new species of Mastodon, and of other vertebrated animals, found on the left bank of the Irrawadi: *Geol. Trans.* 2nd series, II, 369—376, (1829); *Proc. Geol. Soc.*, 69—71, (1834); *Glean. Sci.*, I, 182—184, (1829).

Coal.

1. On the coal-field of Pálamú: *Glean. Sci.*, II, 217—220, (1831).

2. Reports of a Committee for Investigating the Coal and Mineral Resources of India with reference to Coal and Iron. 8°. Calcutta, 1838. Table of assays in *Jour. As. Soc. Beng.*, VII, 197—199, (1838); J. McCLELLAND, No. 10.

3. Reports and abstracts of the Proceedings of a Committee for the Investigation of the Coal and Mineral Resources of India, brought up to May, 1841: Folio. Calcutta, 1841; partly in *Jour. As. Soc. Beng.*, VII, 948—962, (1838), IX, 198—214, (1840).

Coal,—cont.

4. Abstract of the Proceedings of a Committee for the Investigation of the Coal and Mineral Resources of India, brought up to April 1842. Flscp Calcutta, 1842.

5. Notes on Rajmehal Coal (from proceedings of Coal Committee) : *Cal Jour. Nat. Hist.*, III, 501—506, (1843).

6. Analysis of coal from the falls of the Jamuna river : *Cal. Jour. Nat. Hist.* V, 444, (1845).

7. Report of a Committee for the Investigation of the Coal and Mineral Resources of India for May 1845. Folio. Calcutta, 1846.

8. Correspondence concerning the discovery of coal in Upper Assam by Lieutenant Dalton : *Cal. Jour. Nat. Hist.*, VII, 213—216, (1847).

9. Discovery of coal at the foot of the Bhooteah hills : *Cal. Jour. Nat. Hist.*, VIII, 277, (1847).

10. Further discoveries of Coal on the Northern side of Assam : *Cal. Jour. Nat. Hist.*, VII, 368, (1847).

11. Papers on the coal of the Nerbudda Valley, Tenasserim Provinces and Thayetmyo : *Sel. Rec. Gov. Ind.*, X, 1856.

12. Notice of an exploration for Coal near Kotree in Scinde : *Mad. Jour. Lit. Sci.*, XIX (new series, III), 142, (1857). (Reprinted from *Scindian*, 13th January, 1857.)

13. Official correspondence regarding the existence of Coal and Iron in the Punjab. 8°. Calcutta, 1859.

14. Reports on the coal Resources and Production of India : *Sel. Rec. Gov. Ind.*, LXIV, (1868).

15. Papers on the Coal and Iron resources of the Chanda district. Flscp. Calcutta, 1874.

16. Proceedings of a Committee assembled by order of His Excellency the Governor-General-in-Council, under date 28th January, 1874, to consider and report on the best means of improving the steam service on the Brahmaputra and the water communication with Upper Assam : *Assam Gazette*, 17th April, 1875.

Cockburn, J.

1. Notes on Stone Implements from the Khasi hills and the Banda and Vellore Districts : *Jour. As. Soc. Beng.*, XLVIII, pt. ii, 133—143, (1879).

2. Stone implements from Banda : *Proc. As. Soc. Beng.*, 1882, p. 8.

Coddington, F.

1. Description of country surveyed in District Chanda, season 1867-68, [coal and sandstone] : *Sel. Rec. Gov. Ind.*, LXXIV, 103, (1869).

Cole, R.

1. On the Geological position and association of the Laterite, or Iron clay, formation in India, with a description of that Rock as it is found at the Red Hills near Madras : *Mad. Jour. Lit. Sci.*, IV, 100—106, (1836).

2. Note on Campbell's remarks on Dr. Brenza's nomenclature of Indian minerals : *Mad. Jour. Lit. Sci.*, X, 440, (1839).

Colebrooke, H. T.

1. On the valley of the Sutluj River in the Himálaya mountains: *Geol. Trans.*, 2nd series, I, 124—131, (1824).

2. On the Geology of the North-Eastern Border of Bengal: *Geol. Trans.*, 2nd series, I, 132—136, (1824).

Colebrooke, H. T. *and* Gerard, A.

1. On the valley of the Setlej River, in the Himalayan Mountains, from the Journal of Captain A. Gerard; with Remarks by Henry Thomas Colebrooke: *Trans. Roy. As. Soc.*, I, 343—380, (1826); *Edin. Jour. Sci.*, V, 270—278, VI, 28—50, (1826).

Colebrooke, R. H.

1. On the Islands Nancoury and Comarty: *As. Res.*, IV, 129—133, (1795).

2. On the Andaman Islands: *As. Res.*, IV, 385—396, (1795).

3. On Barren Island and its Volcano: *As. Res.*, 397—400, (1795).

4. On the course of the Ganges through Bengal: *As. Res.*, VII, 1—31, (1801).

Collier, C. F.

1. On the nature of the soils of the Bombay Presidency: *Trans. Bo. Geog. Soc.*, IX, 99—110, (1856).

Collingwood, C.

1. Rambles of a naturalist on the shores and waters of the China Sea: being observations in natural history during a voyage to China, Formosa, Borneo, Singapore, &c., made in Her Majesty's vessels in 1866 and 1867. 8°. London, 1868.

Colvin, J.

1. [Discovery of fossil bones in alluvium at Dum Dum]: *Jour. As. Soc. Bengal*, II, 649, (1833).

2. Additional fragments of the Sivatherium: *Jour. As. Soc. Bengal*, VI, 152—153, (1837).

3. Catalogue of a second collection of fossil bones, presented to the Society's Museum: *Jour. As. Soc. Bengal*, V, 179—183, (1836).

Congalton.

1. Examination of the coast of the Peninsula from P. Mutiara to P. Panjang in search of coal deposits in November, 1847: *Jour. Ind. Archip.*, I, 353*—358,* (1847).

Congreve, H.

1. Observations on the altered rocks of the Neilgherries: *Mad. Jour. Lit. Sci.*, XXII (new series, VI), 49—52, (1861).

2. Contributions to the Geology and Mineralogy of the Neilgherry Hills: *Mad. Jour. Lit. Sci.*, XXII (new series, VI), 226—260, (1861).

3. Note on a raised beach at Aden: *Mad. Jour Lit. Sci.*, 1878, pp. 168—170.

Conolly, A.

1. Note on the Samar lake salt and earth: *Jour. As. Soc. Bengal*, V, 799—801, (1836).

Conolly, E.

1. On an aërolite presented to the Society [fell near Oujein, 23rd June, 1838]: *Jour. As. Soc. Bengal,* VIII, 822-823, (1839).

2. Sketch of the Physical Geography of Seistan : *Jour. As. Soc. Beng.,* IX,. 710—725, (1840).

3. Journal kept while travelling in Seistan : *Jour. As. Soc. Beng.,* X, 319— 340, (1841).

Conybeare, W. D.

1. [Notice of the geology of India] : *Brit. Ass. Rep.,* 1832, p. 395 ; *Jour. As. Soc. Beng.,* II, 606-607.

Cook, H.

1. Geological report on a part of Beloochistan : *Trans. Med. Phys. Soc. Bombay,* V, 105—113, (1859).

2. Topographical and geological sketch of the Province of Sarawan, or northern portion of the table-land of Beloochistan : *Trans. Med. Phys. Soc. Bombay,* VI, 1—44, (1860).

3. Topographical and geological sketch of a portion of the Province of; Jhalawan and the eastern division of Mekran : *Trans. Med. Phys. Soc. Bombay,* VI, 45—103, (1860).

4. Geological discoveries in the valley of Kalat and surrounding parts of Beloochistan: *Jour. Bo. As. Soc.,* VI, 184—194, (1862).

5. Extract from a letter regarding the discovery of a fossil elephant's tusk. near Ootaru on the Waie Road: *Jour. Bo. As. Soc.,* VII, p. xlviii,(1863).

Cooke, C. B.

1. [Tin resources of Tenasserim] : *Ind. Economist,* III, 148-149, (1872).

Copland,

1. Account of the Cornelian mines in the neighbourhood of Broach: *Trans. Lit. Soc. Bombay,* I, 289—295, (1815) ; WESTERN INDIA, 491—495, (1857).

2. Sur la mine de cornaline de Barotch entre Bombay et Brouda : *Bull. Soc. Géol. France,* XIII, 669-670, (1856).

Copper.

1. On the copper works at Singhána near Khetri in the Shekhawati country : *Glean. Sci.,* III, 380—384, (1831).

Corbett, A. F.

1. The climate and resources of Upper India, and suggestions for their improvement. 8°. London, (1874).

Costello, C. P.

1. Observations on the Geological Features, &c., of the country in the neighbourhood of Bunnoo and the Sanatorium of Shaikh Boodeen : *Jour. As. Soc. Beng.,* XXXIII, 378—380, (1864).

Cotton, *Sir* Arthur T.

1. Report on the Mahanuddi Delta. Report on the water communication between Calcutta and the Ganges. Memorandum on the proposed Mari Kanvái Tank in Mysore. Iron or water for India. 8°. Madras, 1859.

Coulthard, S.

1. The Trap formation of the Sagor district and of those districts westwards of it as far as Bhopalpur on the banks of the rivei Newas in Omatwara : *As. Res.*, XVIII, 47—81, (1833) ; *Glean. Sci.*, I, 216—218, (1829) ; WESTERN INDIA, 207—230, (1857).

Cowan, L.

1. Memorandum on the produce of the Himalaya Hills. 8°. Lahore, 1860.

Cox, Hiram.

1. An account of the Petroleum Wells in the Burmha Dominions, extracted from Journal of a voyage from Ranghong up the River Erai Wuddey to Amarapoorah, the present capital of the Burmha Empire : *As. Res.*, VI, 127—137, (1799) : *Phil. Mag.*, IX, 226—234, (1801).

Cracroft, W.

1. [Limestone in Sylhet] : *Jour. As. Soc. Beng.*, I, 74, (1832).

2. Smelting of Iron in the Kasya Hills : *Jour. As. Soc. Beng.*, I, 150—151, (1832).

3. [Information respecting the coal of the Khasia hills] : *Jour. As. Soc. Beng.*, I, 250, 252—253, 363, (1832).

4. Notes relative to the collection of some Geological specimens in the Kasia Hills between Assam and Nanklow : *Jour. As. Soc. Beng.*, I, 561, (1832) ; III, 293—296, (1834).

Crawfurd, H.

1. [Smooth water anchorages on the Travancore coast] : *Mad. Jour. Lit. Sci.*, XXII, 133—136, (1861).

Crawfurd, J.

1. Geological observations made on a voyage from Bengal to Siam and Cochin China : *Geol. Trans*, 2nd series, I, 406—408, (1824).

2. Journal of an Embassy from the Governor-General of India to the Court of Ava in 1827. 4°, London, 1829. *2nd edition*, 2 vols., 8°, London. Edinburgh and Dublin. Contains BUCKLAND, No. 1, in Appendix,1 t ed., pp. 78 to 88 ; 2nd ed., II, pp. 143—162.

3. [Account of the Mission to Ava] : *Gazette of India*, 1st March, 1827 *Edin. New Phil. Jour.*, III, 359—370, (1827).

Crespigny, De.

1. Discovery of beds of Lignite under Laterite at Ratnagherry : *Jour. Bo. As. Soc.*, V, 626—628, (1857).

Criper, W. R.

1. Note on some Antimony Deposits in the Maulmain District : *Rec. Geol. Surv. Ind.*, XVIII, 151—153, (1885).

Croockeurt, H.

1. The Tin mines of Malacca : *Jour. Ind. Archip.*, VIII, 112—133, (1854), [translated from *Tidjschrift Voor. Nederl. Ind.*, November, 1851.]

Cullen, W.

1. Notice of the geological features of a route from Madras to Bellary in April and May, 1822 : *Phil. Mag.*, IV, 355—363, 435—443, (1828).

Cullen, W.,—cont.

2. Extract from a letter regarding graphite in Travancore : *Mad. Jour. Lit. Sci.*, XVIII (new series, II), 295—297, (1857).

Cunliffe, C. E.

1. [Fossils at Pondicherry] : *Cal. Jour. Nat. Hist.*, II, 113—115, (1842).

Cunningham, Alex.

1. Abstract Journal of the Route to the sources of the Punjab rivers : *Jour. As. Soc. Beng.*, X, 1—6, 105—115, (1841).

2. Journal of a trip through Kulu and Láhul, to the Chu Mureri Lake in Ladák, during the months of August and September, 1846 : *Jour. As. Soc. Beng.*, XVII, pt. i, 201—230, (1848).

3. Ladak, Physical, Statistical, and Historical, with notices of the surrounding countries. 8°.   London 1854.

4. On stone and timber of the Gwalior Territory : *Prof. Pap. Thomason Eng. Coll., Roorkee*, No. IV, 1854.

Cunningham, Alex., *and* Broome, A.

1. Abstract Journal of the Routes of Lieutenants A. Broome and A. Cunningham to the sources of the Punjab rivers : *Jour. As. Soc. Beng.*, X, 1—6, (1841).

Cunningham, Allan.

1. Memorandum on the Irawadi river, with a monthly Register of its rise and fall from 1856 to 1858, and a measurement of its minimum discharge : *Jour. As. Soc. Beng.*, XXIX, 175—183, (1860).

Cunningham, J. D.

1. Notes on Moorcroft's travels in Ladakh, and on Gerard's account of Kunauar, including a general description of the latter district : *Jour. As. Soc. Beng.*, XIII, 172—253, (1844).

2. On the embankments of rivers, and on the nature of overflowing rivers in Diluvial plains : *Jour. As. Soc. Beng.*, XVIII, 697—702, (1849).

D

Dalton, E. T.

1. Report of a visit to the hills in the neighbourhood of the Sobanshiri river : *Jour. As. Soc. Beng.*, XIV, 250—267, (1845).

2. Coal from Darjmu river, Upper Assam : *Cal. Jour. Nat. His.*, VII, 214—216, (1847).

3. Earthquakes experienced in Assam in the latter end of Jan. 1849 : *Jour. As. Soc. Beng.*, XVIII, 173—174, (1849).

4. Account of a visit to the Jugloo and Seesee rivers, in Upper Assam ; together with a note on the gold fields of that Province, by Major Hannay : *Jour. As. Soc. Beng.*, XXII, 511—521, (1853).

Dalton, E. T. *and* Hannay, S. F.

1. Note on recent investigations regarding the extent and value of the Auriferous deposits of Assam, being abstracts of reports by Capt. E. T. Dalton and Lieut. Col. S. F. Hannay, dated October, 1855 : *Mem. Geol. Surv. Ind.*, I, pt. i, 94—98, (1856).

Dalton, T.

1. Notes of a tour made in 1863—64, in the Tributary Mehals, under the Commissioner of Chota Nagpore, Bonai, Gangpore, Odeypore, and Sirgooja : *Jour. As. Soc. Beng.*, XXXIV, pt. ii, pp. 31, (1865).

Daly, D. D.

1. Metalliferous formations of the [Malay] Peninsula : *Jour. Straits As. Soc.*, No. II, pp. 193—198, (1878).

Damour, A. A.

1. Note sur la Tcheffkinite de la côte du Coromandel : *Bull. Soc. Geol. France*, 2nd series, XIX, 550—552, (1862).

D'Amato, *Père*, Giuseppe.

1. Short description of the mines of precious stones, in the District of Kyatpen, in the kingdom of Ava [Translation] : *Jour. As. Soc. Beng.*, II, 75—76, (1833) ; Burma, pp. 285—287, (1882).

Dana, J. D.

1. [Minerals from W. India] : *Am. Jour. Sci.*, 2nd series, XI, 424, (1851).

Dangerfield, F.

1. [Geology of Central India] : *Sir* J. Malcolm, *Memoir of Central India*, Vol. II, *Appendix* ii, pp. 313—349, (1823) ; partly in Western India, 231—246, (1857).

Danvers, C.

1. Memo. on Indian Coals, with reference to their employment as fuel for Indian Railways and Steamers : *Gazette of India Supplement*, 1868, pp. 84—89.

Danvers, F. C.

1. Spons' information for Colonial Engineers, No. 3, India. 8°. London, 1877.

D'Archiac, É. T. *and* Haime, J.

1. Description des Animaux Fossiles du groupe Nummulitique de l'Inde, 4°. Paris, 1853.

Davidson, T.

1. On some Carboniferous Brachiopoda collected in India by A. Flemming, M.D. and W. Purdon, Esq.. F.G.S. : *Quart. Jour. Geol. Soc.*, XVIII, 25—35, (1862).

2. Note on the Carboniferous Brachiopoda collected by Captain Godwin-Austen : *Quart. Jour. Geol. Soc.*, XX, 387, (1864).

3. Note on some Jurassic and Cretaceous Brachiopoda, collected by Captain Godwin-Austen in the Mustakh hills in Thibet : *Quart. Jour. Geol. Soc.*, XXII, 35—39, (1866).

4. Note on Carboniferous Brachiopoda collected by Captain Godwin-Austen in the valley of Cashmere : *Quart. Jour. Geol. Soc.*, XXII, 39—45, (1866).

Davy, John.

1. A description of Adam's Peak : *Quart. Jour. Sci.*, V, 25—30, (1818).

Davy, John,—cont.

2. Description of certain rocks in the south of Ceylon : *Quart. Jour. Sci.,* V, 233—235, (1818).

3. Chemical examination of some substances used in Ceylon as remedies against the bites of venomous serpents : *Phil. Mag.,* LI, 122—123, (1818) ; *Jour. de Méd.,* I, 299—300, (1818).

4. Analysis of the snake stone : *As. Res.,* XIII, 316—318, (1820) ; *Jour. de Pharm.,* IX, 162—163, (1823).

5. On the geology and mineralogy of Ceylon : *Geol. Trans.,* 1st series, V, 311—327, (1821).

6. Sur les sources chaudes de Ceylan : *Ann. de chemie,* XXIII, 269—272, (1823).

Dawe, W.

1. Memorandum of the progress of sinking a well in the bunds of Chandpur near the foot of the hills : *Jour. As. Soc. Beng.,* VI, 52—54, (1837).

Day, F.

1. Narrikal, or Cochin mud bank : *Mad. Jour. Lit. Sci.,* XXII (new series, VI), 260—264, (1861).

Dean, E.

1. On the strata of the Jumna alluvium, as exemplified in the Rocks and Shoals lately removed from the bed of the River ; and of the sites of the Fossil Bones discovered therein : *Jour. As. Soc. Beng.,* IV, 261—278, (1835).

2. On the Fossil Bones of the Jumna river ; *Jour. As. Soc. Beng.,* IV, 495—499, (1835).

De Blainville, H. D.

1. Sur une tête de chameau fossile dans un grès des Sous-Himalayas : *Comptes Rendus,* III, 528, (1836) ; *Ann. Sci. Nat. Paris,* VI, (Zool.), 317—319, (1836).

2. Sur le chameau fossile, et sur le Sivatherium des Sous-Himalayas méridionaux : *Comptes Rendus,* IV, 71—76, (1837).

3. Observations sur une note par M. Geoffroy sur le chameau et le Sivatherium fossiles : *Comptes Rendus,* IV, 166—168, (1837).

De Crespigny.

1. Discovery of beds of Lignite under Laterite at Ratnagherry : *Jour. Bo. As. Soc.,* V, 626—628, (1857).

Dejoux, P.

1. On Margohi cement : *Prof. Pap. Ind. Eng.,* 2nd series, II, 12—30, (1873).

Dejoux, P., *and* **Brownlow, H. A.**

1. Notes on the proposed manufacture of Hydraulic Cements in India : *Prof. Pap. Ind. Eng.,* 2nd series, I, 604—621, (1872).

De Königk, L.

1. Descriptions of some fossils from India, discovered by Dr. A. Fleming of Edinburgh [in the Punjab Salt Range] : *Quart. Jour. Geol. Soc.,* XIX, 1—19, (1863).

34

De La Croix, M. J. E.
1. Les mines d'Etain de Perak : *Archives des Missions Scientifiques et Littéraires*, 3rd series; IX, 1—78, (1882).

De La Hoste, E. P.
1. On the Nerbudda river : *Trans. Bo. Geog. Soc.*, I, 174—177, (1838).

2. Memoranda respecting the existence of copper in the territory of Luz, near Beyla: *Jour. As. Soc. Beng.*, IX, 30—32, (1840) ; *Trans. Bo. Geog. Soc.*, VI, 117—119, (1840).

3. Report on the country between Kurrachee, Tatta, and Sehwan, Scinde : *Jour. As. Soc. Beng.*, IX, 907—915, (1840).

De la Rue, W., *and* **Müller, H.**
1. Chemical examination of Burmese Naphtha or Rangoon Tar : *Proc. Roy. Soc. Lond.*, VIII, 221—228, (1857); *Phil. Mag.*, 4th ser., XIII, 512—517, (1857).

Des Mazures.
1. Sur quelques coquilles fossiles du Thibet. Lettre à M. Elie de Beaumont. Détermination de ces fossiles par M. Guyerdet: *Comptes Rendus*, LVIII, 878, (1864) ; *Geol. Mag.*, Decade I, 76, (1864).

Desgodins.
1. Notes géologiques sur la route de Yer-Ka-Lo à Pa-tang : *Bull. Soc. Géogr. Paris*, 1876, pp. 492—508.

De Zigno, Achille.
1. Flora Fossilis Formationis Oolithicæ. Le piante fossili dell' oolite. 4° Padova, (1856—1881).

2. Sopra i Depositi di Piante Fossili dell' America Settentrionale, delle Indie, e dell' Australia che alcuni autori riferiono all' Epoca oolitica: *Geol. Mag.* I, 1st decade, 166, (1864), [quoting *Revista Period. Accad. Sci. Padova.*]

Dickens, C. H.
1. Memorandum of experiments on, and analysis of, specimens of Kunkur from about the 393rd mile-stone on the Grand Trunk road, near Naubatpore: *Sel. Pub. Corr. N.-W. P.*, No. XXXVI, 105—107, (1849); *Sel. Rec. Gov. N.-W. P.*, new series, III, 349—352, (1867).

Dixon, C. G.
1. Some account of the lead mines of Ajmir : *Glean. Sci.*, III, 111—115, (1831).

Dodd, C. D.
1. Particulars concerning the Runn of Cutch and the country on its southern margin : *Trans. Bo. Geog. Soc.*, XVI, 1—7, (1863).

2. Memorandum on the eastern portion of Cutch, called Wagur : *Trans. Bo. Geog. Soc.*, XVI, 7—9, (1863).

Donaldson, J.
1. Report on the utilisation of iron-making materials in the neighbourhood of Hazareebagh by means of convict labour. Flscp. Dum Dum, (1870).

Douglas, C.

1. Report on the River Jumna between Agra and Ooreah: *Sel. Rec. Gov. N.-W. P.*, new series, IV, 437—456, (1868).

Doyle, P.

1. On some Tin Deposits of the Malay Peninsula: *Quart. Jour. Geol. Soc.*, XXXV, 229—232, (1879).

2. Tin mining in Larut. 8°. London, 1879.

3. A contribution to Burman Mineralogy. 8°. Calcutta, 1879.

4. Prospect of Artesian borings in the Bellary district. 8°. Madras, 1883.

5. Coal mining by Blasting in the Bengal coal-field: *Prof. Pap. Ind. Eng.*, 3rd series, III, 157—159, (1885).

Drew, F.

1. Alluvial and Lacustrine deposits and Glacial Records of the Upper Indus Basin, Pt. I, Alluvial deposits: *Quart. Jour. Geol. Soc.*, XXIX, 441—471, (1873).

2. [Reply to Col. Greenwood's remarks on the above paper]: *Geol. Mag.*, 2nd decade, I, 94, (1874).

3. The Jummoo and Kashmir territories. A geographical account. 8°. London, 1875.

Drummond, H.

1. Report on the copper mines of Kumaon: *Jour. As. Soc. Beng.*, VII, 934—940, (1838).

2. On the Mines and Mineral Resources of Northern Afghanistan: *Jour. As. Soc. Beng.*, X, 74—93, (1841).

3. Report on the deposits of Graphite, near Almorah: *Sel. Pub. Corr. N.-W. P.*, No. L, 213—218, (1850); *Sel. Rec. Gov. N.-W. P.*, new series, III, 371—378, (1867).

4. Report on the iron of the Province of Kumaon and Gurhwal: *Sel. Rec. Gov. Ind.*, VIII, *Supplement*, pp. 20—33, (1855).

Drury, Heber.

1. Notes of an excursion along the Travancore Backwater: *Mad. Jour. Lit. Sci.*, XIX, 203—219, (1857).

Duckworth, H.

1. On the Fossils of Perim Island, in the gulf of Cambay: *Proc. Liverpool Geol. Soc.*, 1859—69, p. 94; *Proc. Liverpool Lit. Phil. Soc.*, XII, 142—157, (1857-58).

2. Description of part of the lower jaw of a large mammal, probably of Deinotherian type, from Perim Island, Gulf of Cambay, India: *Proc. Liverpool Geol. Soc.*, VI, 38—40, (1865).

Duff, A.

1. Account of the Nat Mee, or the Spirit fire, a burning hillock in the Province of Pegu: *Jour. As. Soc. Beng.*, XXX, 309—313, (1861).

Dunbar, W.

1. Discovery of coal in a new site [at Balia, Hazaribagh]: *Jour. As. Soc. Beng*, X, 300—301, (1841).

Duncan, P. M.

1. A description of, and remarks upon, some Fossil Corals from Scinde: *Quart. Jour. Geol. Soc.*, XX, 66—72, (1864); *Ann. Mag. Nat. Hist.*, XIII, 3rd series, 295—307, (1864).

2. A description of the Echinodermata from the Strata of the South Eastern coast of Arabia, and at Bagh on the Nerbudda, in the collection of the Geological Society: *Quart. Jour. Geol. Soc.*, XXI, 349—363, (1865)

3. An abstract of the geology of India. Folio. London, 1875; 2nd edition, 1876.

4. Scientific Results of the second Yarkand Mission. Syringosphæridæ. 4°. Calcutta, 1879.

5. Sind Fossil corals and Alcyonaria: *Pal. Indica*, series vii, xiv, I, pt. ii, (1880.)

6. On the Coralliferous series of Sind and its connection with the last upheaval of the Himalayas: *Quart. Jour. Geol. Soc.*, XXXVII, 190—209, (1881).

7. On the Echinoidea of the Cretaceous Strata of the Lower Nerbudda Region: *Quart. Jour. Geol. Soc.*, XLIII, 150—155, (1887).

8. Note on the Echinoidea of the cretaceous series of the Lower Nerbada Valley, with remarks upon their Geological age: *Rec. Geol. Surv. India*, XX, 81—92, (1887).

Duncan, P. M. *and* Sladen, W. P.

1. Fossil Echinoidea of Western Sind and the coast of Biluchistan and of the Persian Gulf, from the Tertiary formation: *Pal. Indica*, series vii, xiv, I, pt. iii, (1882-86).

2. The Fossil Echinoidea of Kuch and Kattywar: *Pal. Indica*, series vii, xiv, I, pt. iv, (1883).

Durand, H. M.

1. Specimens of the Hippopotamus and other Fossil Genera of the Sub-Himalayas in the Dádúpur Collection: *As. Res.*, XIX, pt. i, 54—59, (1836).

2. [Extract from letter to M. de Blainville accompanying a skull of a fossil camel from the Sivaliks:] *Comptes Rendus*, III, 529, (1836).

Durand, H. M. *and* Baker, W. E.

1. Table of Sub-Himalayan Fossil Genera in the Dádúpur Collection: *Jour. As. Soc. Beng.*, V, 291—293, 486—504, 661—669, 739—740, (1836); Fossil monkey's jaw in *Ann. Sci. Nat. Paris*, VII, (Zool.), 370—372, (1837); *Phil. Mag* XI, 33—36, (1837); *Edin. New Phil. Jour.*, XXIII, 216—217, (1837).

2. Fossil remains of the smaller Carnivora from the Sub-Himalayas: *Jour. As. Soc. Beng.*, V, 576—584, (1836).

Durrschmidt, C.

1. Report on the copper mines of Singbhoom, S.-W. Frontier of Bengal, 8°. Calcutta, 1857.

Dykes, J. W.

1. On the increase of land on the Coromandel coast (abstract): *Quart. Jour. Geol. Soc.*, XVII, 533, (1861).

E

E, A.
1. Iron works at Ferozepur : *Glean. Sci.*, III, 327—328, (1831).
2. On the copper works at Singhana, near Khetri, in the Shekhawâti country : *Glean. Sci.*, III, 380—384.

East, *Sir* Edward Hyde.
1. Abstract of an account containing the particulars of a boring made near the River Hooghly, in the vicinity of Calcutta, from May to July, 1814, inclusive, in search of a spring of pure water : *As. Res.*, XII, 542—546, (1816).

Earthquake.
1. Account of the Earthquake at Kutch on the 16th June, 1819, drawn up from published and unpublished letters from India : *Edin. Phil. Jour.*, III, 120—124, (1820).
2. Earthquake of the 26th August 1833 : *Jour. As. Soc. Beng.*, II, 438—439, (1833).
3. Notice of an earthquake felt in Sind, 28th October, 1870 : *Proc. As. Soc. Beng.*, 1871, p. 56.
4. Record of the occurrence of Earthquakes in Assam, during the years 1874—80 : *Jour. As. Soc. Beng.*, XLVI, pt. ii, 294, (1877) ; XLVII, pt. ii, 48, (1878) ; L, pt. ii, 61, (1881).

Economic Geology.
1 Correspondence respecting the Society's Museum of economic geology : *Jour. As. Soc. Beng.*, XI, 326—340, (1842).

Egerton, *Sir* P. de Malpas Grey.
1. On the remains of fishes found by Mr. Kaye and Mr. Cunliffe in the Pondicherry beds : *Quart. Jour. Geol. Soc.*, I, 164—171, (1845); *Geol. Trans.*, 2nd series, VII, 89—96, (1846).
2. Description of the specimens of Fossil Fish from the Deccan, India : *Quart. Jour. Geol. Soc.*, VII, 273, (1851) ; Western India, 302, (1857).
3. On two new species of Lepidotus from the Deccan : *Quart. Jour. Geol. Soc.*, X, 371—374, (1854).
4. On an Ichthyoid Fossil from India : *Brit. Ass. Rep.*, 1854, pt. ii, p. 82.

Egerton, *Sir* P. de Malpas Grey, *and* L. C. Miall.
1. The Vertebrate fossils of the Kota-Maleri group : *Pal. Indica*, 4th series, I, pt. 2, (1878).

Egerton, R. E.
1. [Supposed effects of the earthquake in Murwut in Bunnoo in moistening the soil] : *Proc. As. Soc. Beng.*, 1869, pp. 163—164.

Ehrenberg, C. G.
1. Ausgedehnte Felsbildung aus Kieselschaligen Polycystinen auf den Nicobaren Inseln : *Monatsber. K. Akad. Wiss. Berlin*, 1850, pp. 476—478.
2. Mikrogeologie. Das Erden und Felsen schaffende Wirken des unsichtbar kleinen selbständigen Lebens auf der Erde. Folio. Leipzig, 1854—56.

Ehrenberg, C. G.,—cont.

3. Fortsetzung der mikrogeologischen Studien, &c.; Poly-cystinen Gebirge der Nicobaren Inseln: *Abhandl. K. Akad. Wiss., Berlin,* 1875, pp. 116—120.

Ehrenberg, E. G.

1. On the nature and formation of coral Islands and coral Banks in the Red Sea: *Jour. Bo. As. Soc.,* I, 73—83, (1841); 129—136, (1842); 322—341, (1843); 390—402, (1844).

Elliot, W.

1. [Note on the intertrappean beds and fossils near Rajahmundry]: *Jour. As. Soc. Beng.,* XXIII, 309—310, 399—400, (1854).

Elphinstone, *Hon'ble* **Mountstuart.**

1. Account of the Kingdom of Caubul and its Dependencies. 4°. London: (1815).

Etheridge, R.

1. Note on the Jurassic Fossils collected by Captain Godwin-Austen [in Kashmir]: *Quart. Jour. Geol. Soc.,* XX, 387—388, (1864).

Ethersey, R.

1. Note on Perim Island in the Gulf of Cambay: *Trans. Bo. Geog. Soc.,* Nov., 1838, pp. 54—58; WESTERN INDIA, 472—474, (1857).

Evans, John.

1. On some flint cores from the Indus, Upper Scinde: *Geol. Mag.,* III, 433—435, (1866).

Evans *and* **R. H. Keatinge.**

1. Report on a passage made on the Nurbudda River, from the falls of Dharee to Mundlaisir by Lieut. Keatinge, and of a similar passage from Mundlaisir to Baroach by Lieutenant Evans: *Jour. As. Soc. Beng.,* XVI, 1104—1112, (1847).

Everest, R.

1. Geological observations made on a journey from Calcutta to Ghazipur: *Glean. Sci.,* III, 129—136, (1831).

2. On the Sandstone of India: *Glean. Sci.,* III, 207—213, (1831).

3. Note on Indian Saline Deposits. [Bundelkhund and Bhartpore]: *Jour. As. Soc. Beng.,* I, 149—150, (1832).

4. Some observations on the quantity of earthy matter brought down by the Ganges River: *Jour. As. Soc. Beng.,* I, 238—241, 549—550, (1832).

5. Remarks on a late paper in the Asiatic Journal on the Gypsum of the Himalaya: *Jour. As. Soc. Beng.,* I, 450—454, (1832).

6. Memorandum on the Fossil Shells discovered in the Himalayan Mountains: *As. Res.,* XVIII, pt. ii., 107—114, (1833).

7. Some Geological remarks made in the country between Mírzapúr and Ságar, and from Ságar northwards to the Jumna: *Jour. As. Soc. Beng.,* II, 475—481, (1833).

8. On the climate of the fossil elephant: *Jour. As. Soc. Beng.,* III, 18—24, (1834).

Everest, R.,—cont.

9. On the temperature of deep wells to the West of the Jumna: *Jour. As. Soc. Beng.*, IV, 229, (1835); *Bibl. Univ.*, IV, 355—356, (1836).

10. Geological observations made in a journey from Musooree (Masúri) to Gungotree (Gangautri): *Jour. As. Soc. Beng.*, IV, 692—693, (1835); *Bibl. Univ.*, V, 410—412, (1836).

11. Some Geological remarks made in a journey from Delhi, through the Himalayan mountains, to the frontier of Little Thibet, during 1837: *Proc. Geol. Soc.*, III, 566—570, (1841).

12. On the high temperature of wells in the neighbourhood of Delhi: *Proc. Geol. Soc.*, III, 732—735, (1842).

F

F.

1. Some account of the Casiah Hills: *Glean. Sci.* I, 252—255, (1829).

Falconer, H.

1. Note on certain specimens of Animal Remains from Ava, presented by James Calder, Esq., to the Museum of the Asiatic Society: *Glean. Sci.*, III, 167—170, (1831).

2. [Letter on the Dehra Dun fossils]: *Jour. As. Soc. Beng.*, I, 249, (1832).

3. On the aptitude of the Himálayan Range for the Culture of the Tea Plant: *Jour. As. Soc. Beng.*, III, 178—188, (1834).

4. [Letter on Siwalik fossils]: *Jour. As. Soc. Beng.*, IV, 57—59, (1835).

5. Note on the occurrence of fossil bones in the Sewalik Range, eastward of Hardwar: *Jour. As. Soc. Beng.*, VI, 233, (1837).

6. [Account of a very extraordinary elastic sandstone]: *Jour. As. Soc. Beng.*, VI, 240—241, (1837).

7. [Letter to the Secretary of the Asiatic Society on the Cataclysm of the Indus]: *Jour. As. Soc. Beng.*, X, 615—619, (1841).

8. Description of some fossil remains of Dinotherium, Giraffe, and other Mammalia from the Gulf of Cambay, in India: *Quart. Jour. Geol. Soc.*, I, 356—372, (1845); WESTERN INDIA, 475—490, (1857).

9. Abstract of a discourse on the fossil Fauna of the Sewalik hills: *Jour. Roy. As. Soc.*, VIII, 107—112, (1846).

10. On the asserted occurrence of Human bones in the ancient fluviatile deposits of the Nile and the Ganges, with comparative remarks on the alluvial formation of the two valleys: *Quart. Jour. Geol. Soc.*, XXI, 372—389, (1865).

11. Palæontological Memoirs and Notes. With a biographical sketch. Edited by Charles Murchison. 2 vols., 8°. London, 1868.

Falconer, H., *and* Cautley, *Sir* P. T.

1. Synopsis of fossil genera and species from the upper deposits of the tertiary strata of the Siválik hills, in the collection of the authors: *Jour. As. Soc. Beng.*, IV, 706—707, (1835).

Falconer, H. *and* **Cautley,** *Sir* **P. T.,**—cont.

2. Sivatherium giganteum, a new Fossil Ruminant genus from the valley of the Markanda, in the Siválik Branch of the Sub-Himalayan Mountains : *As. Res.*, XIX, pt. i., 1—24, (1836) ; *Jour. As. Soc. Beng.*, V, 35—50, (1836) ; *Ann. Nat. Sci. Paris, (Zool.)*, V, 348—370, (1836) ; *Bibl. Univ.*, IV, 392—397, (1836).

3. Note on the Fossil Hippopotamus of the Siválik hills : *As. Res.*, XIX, pt. i, 39—53, (1836).

4. Note on the Fossil Camel of the Siválik Hills : *As. Res.*, XIX, pt. i, 115 —142, (1836).

5. Note on the Felis Cristata, a new Fossil tiger, from the Sewálik Hills : *As. Res.*, XIX, pt. i, 135—192, (1836).

6. Note on the Ursus Sivalensis, a new Fossil species from the Siválik Hills : *As. Res.*, XIX, pt. i, 193—201, (1836).

7. Notice on the Remains of a Fossil Monkey from the Tertiary Strata of the Sewalik Hills in the North of Hindoostan : *Geol. Trans.*, 2nd series, V, 499—504, (1840) ; *Proc. Geol. Soc.*, II, 568—569, (1838) ; *Mad. Jour. Lit. Sci.*, XII, 304—309, (1840).

8. On additional Fossil species of the Order Quadrumana from the Sewálik Hills : *Jour. As. Soc. Beng.*, VI, 354—360, (1837).

9. Communication on the Colossochelys Atlas : *Proc. Zool. Soc.*, XII, 54—55, 84—88, (1844).

10. On some Fossil Remains of Anoplotherium and Giraffe from the Sewalik Hills in the north of India : *Proc. Geol. Soc.*, IV, 235—249, (1846) ; *Ann. Mag. Nat. Hist.*, 1st series, XIV, 501—502, (1844) ; XV, 55—59, (1845) ; *Cal. Jour. Nat. Hist.*, V, 577—589, (1845).

11. Fauna Antiqua Sivalensis, being the Fossil Zoology of the Siwalik Hills in the north of India. 8°, with folio Atlas. London, 1846.

Fayrer, *Sir* **J.**

1. [Mineral resources and Geological Survey of India] : *Jour. Soc. Arts,* XXX, 595, (1882).

Fedden, F.

1. Report on the nature of the country passed through by the Expedition to the Salween, and the result of observations at the River as to its Navigability, with Meteorological tables and a route map : *Sel. Rec. Gov. Ind.*, XLIX, 30—81, (1865).

2. On the evidences of "ground-ice" in tropical India during the Talchir period : *Rec. Geol. Surv. Ind.*, VIII, 16—18, (1875).

3. Distribution of the fossils described by Messrs. D'Archiac and Haime in the different tertiary and infra-tertiary groups of Sind : *Mem. Geol. Surv. Ind.*, XVII, 197—210, (1879).

4. Popular Guide to the Geological collections in the Indian Museum, Calcutta, No. 3, Meteorites. 8°. Calcutta, 1880.

5. The Geology of the Káthiáwar Peninsula in Guzerat : *Mem. Geol. Surv. Ind.*, XXI, 73—136, (1884).

Feistmantel, O.

1. [Stoliczka's Yarkand Fossils and age of the Gondwána rocks; letter to Baron Richthofen] : *Zeits. Deutsch. Geol. Gesel.*, XXVII, 945—949, (1875).

2. Fossile Pflanzen aus Indien. [Letter to Hofrath von Hauer] : *Verhandl. K. K. geol. Reichs. Wien.*, 1875, pp. 187-194.

3. Weitere Bemerkungen über fossile Pflanzen aus Indien : *Verh. K. K. Geol. Reichs. Wien.*, 1875, pp. 252—261.

4. Mineralogische notizen aus Indien : *Verhandl. K. K. Geol. Reichs. Wien.*, 1875, pp. 301—303.

5. Flora of the Kach (Cutch) series : *Rec. Geol. Surv. Ind.*, IX, 29—34, (1876).

6. Flora of the Rajmehal series : *Rec. Geol. Surv. Ind.*, IX, 34—42, (1876).

7. Flora and probable age of the Panchet group : *Rec. Geol. Surv. Ind.*, IX, 65—67, (1876).

8. Flora and probable age of the Damuda formation : *Rec. Geol. Surv. Ind.*, IX, 67—78, (1876).

9. Fossil flora of the Talchirs : *Rec. Geol. Surv. Ind.*, IX, 78—79, (1876).

10. On the homotaxis of the Gondwána system : *Rec. Geol. Surv. Ind.*, IX, 115—125, (1876).

11. Flora of the Jabalpur group : *Rec. Geol. Surv. Ind.*, IX, 125—135, (1876).

12. Description of some new, and discussions on some already known, species from the Gondwana system : *Rec. Geol. Surv. Ind.*, IX, 135—144, (1876).

13. Contributions towards the knowledge of the Indian Fossil Flora. On some Fossil Plants from the Damuda series in the Raniganj coal-fields, collected by Mr. J. Wood-Mason : *Jour. As. Soc. Beng.*, XLV, pt. ii, 329—382, (1876) ; *Proc. As. Soc. Beng.*, 1876, pp. 223—228.

14. The Jurassic Flora of Kach : *Pal. Indica*, series ii, xl, xii, II,—pt. i, (1876).

15. Weitere Bemerkungen über die Pflanzen-führenden schichten in Indien und deren mögliches Alter : *Verhandl. K. K. Geol. Reichs. Wien.*, VIII, 165—168, (1876).

16. Ueber die indisschen Cycadeengattungen *Ptilophyllum* Morr. und *Dictyozamites* Old ; *Palæontographica*, 1876, pp. 1—24.

17. Ueber die gattung *Williamsonia* : *Palæontographica*, 1877, pp. 25—51.

18. On the Gondwana Series of India as a probable Representative of the Juro-Triassic epoch of Europe : *Geol. Mag.*, 2nd decade, III, 481—491, (1876) ; IV, 188, (1877).

19. Some Fossil plants from the Atgarh Sandstones : *Rec. Geol. Surv. Ind.*, X, 68—70, (1877).

20. On true *Pterophyllum* from the Raniganj field, and the Cycadeaceæ from the Damuda series : *Rec. Geol. Surv. Ind.*, X, 70—73, (1877).

21. Plant fossils from the Barákar District (Barákar group) : *Rec. Geol. Surv. Ind.*, X, 73—75, (1877).

22. Fossil plants from near Assensole (Raniganj group) : *Rec. Geol. Surv. Ind.*, X, 75, (1877).

Feistmantel, O.,—cont.

23. Explanatory note on *Glossopteris* and *Gangamopteris*: *Rec. Geol. Surv. Ind.*, X, 76, (1877).

24. On a tree-fern stem from the Cretaceous rocks near Trichinopoli in Southern India: *Rec. Geol. Surv. Ind.*, X, 133—137, (1877).

25. Notes on the Karharbári Flora: *Rec. Geol. Surv. Ind.*, X, 137—139, (1877).

26. On the occurrence of *Glossopteris* in the Panchet group and in the Upper Gondwánas: *Rec. Geol. Surv. Ind.*, X, 139—140, (1877).

27. Some elements of the Arctic and Siberian jurassic Floras amongst the plants of the Gondwana system: *Rec. Geol. Surv. Ind.*, X, 196—199, (1877).

28. Notes on *Vertebraria, Schizoneura, Zeugophyllites* and *Nœggerathia*: *Rec. Geol. Surv. Ind.*, X, 199—203, (1877).

29. On the Occurrence of the Cretaceous genus *Omphalia* near Namcho Lake, Tibet: *Rec. Geol. Surv. Ind.*, X, 21—26, (1877).

30. Note on *Estheria* in the Gondwana formation: *Rec. Geol. Surv. Ind.*, X, 26—30, (1877).

31. Note on "*Eryon*, comp. *Barrovensis*," McCoy, from the Sripermatur group, near Madras: *Rec. Geol. Surv. Ind.*, X, 193—196, (1877).

32. On Giant's Kettles (pot-holes) caused by water-action in streams in the Rajmahal Hills and Barakur district: *Proc. As. Soc. Beng.*, 1877, pp. 77—80.

33. Preliminary report and notes on the flora of the Kurhurbali Coal-field, its importance and relations to that of the Talchir shales and Damuda series. 8°. Calcutta, 1877. (Printed privately.)

34. Kurze Bemerkungen über das Alter der sog. alteren Kohlenführenden Schichten in Indien: *Neu. Jahrb. Min. Geol.*, 1877, pp. 147—159, 809—811.

35. [Giants' Cauldrons]: *Neu. Jahrb. Min. Geol.*, 1877, pp. 509—511.

36. Jurassic (Liassic) Flora of the Rajmahal group in the Rajmahal Hills: *Pal. Indica*, series ii, xi, & xii, Vol. I, pt. ii, (1877).

37. Jurassic (Liassic) Flora of the Rajmahal group from Golapilli near Ellore, South Godavari: *Pal. Indica*, series ii, xi & xii, Vol. I, pt. ii, (1877).

38. Flora of the Jabalpur group: *Pal. Indica*, series ii, xi & xii, Vol. II, pt. ii, (1878).

39. Palæontological notes from the Satpura coal basin: *Rec. Geol. Surv. Ind.*, XII, 74—83, (1879).

40. Notes on the genus *Sphenophyllum* and other *Equisetaceæ*, with reference to the Indian form *Trizygia speciosa*: *Rec. Geol. Surv. Ind.*, XII, 163—166, (1879).

41. Upper Gondwana flora of the Outliers on the Madras coast: *Pal. Indica*, series ii, xi & xii, Vol. I, pt. iv, (1879).

42. The Flora of the Talchir Karharbari beds: *Pal. Indica*, series ii, xi, & xii, Vol. III, pt. i, (1879); Supplement, (1881).

43. [Letter on age of Gondwanas]: *Neu. Jahrb. Min. Geol.*, 1879, pp. 58—62.

Feistmantel, O.,—cont.

44. Note on the fossil genera *Nœggerathia*, Stbg., *Nœggerathiopsis*, Fstm., and *Rhiptozamites*, Schmalh., in the Palæozoic and secondary rocks of Europe, Asia, and Australia : *Rec. Geol. Surv. Ind.*, XIII, 61—62, (1880).

45. Notes on fossil plants from Kattywar, Shekh Budin, and Sirgujah : *Rec. Geol. Surv. Ind.*, XIII, 62—69, (1880).

46. Palæontological notes from the Karharbari and S. Rewa fields : *Rec. Geol. Surv. Ind.*, XIII, 176—190, (1880).

47. Further notes on the correlation of the Gondwana flora with other floras : *Rec. Geol. Surv. Ind.*, XIII, 190—193, (1880).

48. Further notes on the correlation of the Gondwana flora with that of the Australian coal-bearing system : *Rec. Geol. Surv. Ind.*, XIII, 250—253, (1880).

49. Flora of the Damuda and Panchet Divisions : *Pal. Indica*, series ii, xi & xii, III, pt. ii, (1880) ; pt. iii, (1881).

50. Notes on some Rajmahal plants : *Rec. Geol. Surv. Ind.*, XIV, 148—152, (1881).

51. Palæontological notes from Hazaribagh and Lohardagga Districts : *Rec. Geol. Surv. Ind.*, XIV, 241—263, (1881).

52. A sketch of the history of the fossils of the Indian Gondwana system : *Jour. As. Soc. Beng.*, L, pt. ii, 168—219, (1881).

53. Popular guide to the Geological Collections in the Indian Museum, Calcutta, No. 4, Palæontological collections. 8°. Calcutta, 1881.

54. Note on remains of palm leaves from the Murree and Kasaoli (Tertiary) beds : *Rec. Geol. Surv. Ind.*, XV, 51—53, (1882).

55. Fossil Flora of the South Rewah Gondwana basin : *Pal. Indica*, series ii, xi & xii, IV, pt. i, (1882).

56. Palæontological notes from the Daltonganj and Hutar coalfields in Chota Nagpur : *Rec. Geol. Surv. Ind.*, XVI, 175—178, (1883).

57. The fossil flora of some of the coal-fields in Western Bengal : *Pal. Indica*, series ii, xi & xii, IV, pt. ii, (1886).

58. Ueber die pflanzen und kohlen-führenden Schichten in Indien (beziehungsw. Asien), Afrika und Australien und darin vorkommende glaciale Enscheinungen : *Sitz. K. Böhm. Gesel. Wis.*, 1887, pp. 1—109.

Fenwick, R. H.

1. [Letters and Reports on Nerbudda Valley coal] : *Sel. Rec. Bo. Gov.*, XIV, 74—101, (1854).

Fergusson, J.

1. On recent changes in the delta of the Ganges : *Quart. Jour. Geol. Soc.*, XIX, 321—354, (1863).

Fife, J. G.

1. Report on the Eastern Narra : *Sel. Rec. Bo. Gov.*, new series, LX, 1861.

Finnis, J.

1. Notice of coal near Hoshungabad: *Glean. Sci.*, III, 293—294, (1831).

2. A Summary description of the geology of the country between Hoshungabad, on the Nerbudda, and Nagpoor, by the direction of Baitool: *Jour. As. Soc. Beng.*, III, 71—75, (1834); Western India, 467—471, (1857).

Fisher, F. H.

1. Geological sketch of Masúri and Landour, in the Himalayas; together with an abstract of the Thermometrical Register kept at Landour, during the year 1831: *Jour. As. Soc. Beng.*, I, 193—195, (1832).

2. Memoir of Sylhet, Kachar, and the adjacent districts: *Jour. As. Soc. Beng.*, IX, 808—848, (1840).

Flemming, A.

1. Report on the Salt Range, and on its Coal and other Minerals: *Jour. As. Soc. Beng.*, XVII, pt. ii, 500—526, (1848).

2. Diary of a Trip to Pind Dadun Khan and the Salt Range: *Jour As. Soc. Beng.*, XVIII, 661—693, (1849).

3. On the Salt Range of the Punjáb [Extracts of letters]: *Quart. Jour. Geol. Soc.*, IX, 189—200, (1853).

4. On the Geology of the Sooliman Range: *Quart. Jour. Geol. Soc.*, IX, 346—349, (1853); Western India, 528—530, (1857).

5. Report on the Geological Structure and Mineral Wealth of the Salt Range in the Punjab: *Jour. As. Soc. Beng.*, XXII, 229—279, 338—368, 444—462, (1853); *Sel. Pub. Corr. Punjab*, II, No. x, pp. 253—381, (1854).

6. Notes on the Iron ore of Koróna in the Jetch Dooab of the Punjab, with a qualitative analysis of the same: *Jour. As. Soc. Beng.*, XXIII, 92—94, (1854).

Foley, W.

1. On coal from Arracan: *Jour. As. Soc. Beng.*, II, 368, (1833).

2. [On Fossil Shells and Coal from Kyouk Phyoo, Ramree]: *Jour. As. Soc. Beng.*, III, 412, (1834).

3. Journal of a tour through the Island of Ramree, with a geological sketch of the country and a brief account of the customs, &c., of its inhabitants: *Jour. As. Soc. Beng.*, IV, 20—38, 82—94, 199—2c6, (1835).

4. Notes on the Geology, &c., of the country in the neighbourhood of Maulmayeng, *vulg.* Moulmein: *Jour. As. Soc. Beng.*, V, 269—280, (1836).

Foote, R. B.

1. Notes of a recent excursion to the Kolymullays: *Mad. Quart. Jour. Med. Sci.*, new series, IV, 91—97, (1862).

2. On the occurrence of Stone Implements in various parts of the Madras and North Arcot Districts, with illustrations: *Mad. Jour. Lit. Sci.*, 3rd series, pt. II, pp. 1—35, (1866).

3. On the Distribution of Stone Implements in Southern India: *Quart. Jour. Geol. Soc.*, XXIV, 484—494, (1868).

Foote, R. B.,—cont.

4. Notes on the geology of the neighbourhood, Madras: *Rec. Geol. Surv. Ind.*, III, 11—17, (1870); *Chingleput Manual.* 8°. Madras, 1879, pp. 5—8 and 383—390.

5. Enquiry into an alleged discovery of coal near Gooty, and of the indications of coal in the Cuddapah District: *Rec. Geol. Surv. Ind.*, IV, 16—18, (1871).

6. Notes on the geology of the country between the towns of Juggiapett and Bellamkonta in the Kistna District: *Mem. Geol. Surv. Ind.*, VIII, 293—313, (1872).

7. On the discovery of Prehistoric Remains in India (Bellary): *Geol. Mag.*, 1st decade, X, 187, (1873).

8. On the geology of parts of the Madras and North Arcot districts lying north of the Palar river, and included in the sheet 78 of the Indian Atlas: *Mem. Geol. Surv. Ind.*, X, 1—32, (1873).

9. The auriferous rocks on the Dambal Hills, Dhárwar District: *Rec. Geol. Surv. Ind.*, VII, 133—142, (1874).

10. Rhinoceros deccanensis: *Pal. Indica*, series x, I, p. 1, (1879).

11. The geological features of the South Mahratta country and adjoining districts: *Mem. Geol. Surv. Ind.*, XII, pt. i, 1—268, (1876).

12. [Geology of] Ratnágiri: *Bombay Gazetteer*, X, 12—21, (1880); BOMBAY 28—37.

13. [Geology of the] Southern Mahratta country: BOMBAY, 53—67.

14. [Geology of] Sáwantwári: *Bombay Gazetteer*, X, 390—399, (1880); BOMBAY, 38—49.

15. Notes on representatives of the Upper Gondwana series in Trichinopoly and Nellore Kistna districts: *Rec. Geol. Surv. Ind.*, XI, 247—259, (1878).

16. On the geological structure of the Eastern Coast from latitude 15° northward to Masulipatam: *Mem. Geol. Surv. Ind.*, XVI, 1—107, (1879).

17. On the Geology of N. Madura, Pudukotai State and S. Tanjore, and Trichinopoly within the limits of sheet 80: *Rec. Geol. Surv. Ind.*, XII, 141—158, (1879).

18. Rough notes on the Cretaceous Fossils from Trichinopoly: *Rec. Geol. Surv. Ind.*, XII, 159—162, (1879).

19. Sketch of the Geology of the N. Arcot District: *Rec. Geol. Surv. Ind.*, XII, 187—208, (1879).

20. [Review of the] Manual of the Geology of India: *Geol. Mag.*, Dec. ii, VII, 79—85, 127—134, (1880).

21. Sketch of the work of the Geological Survey in Southern India: *Mad. Jour. Lit. Sci.*, 1881, pp. 279—328.

22. Notes on the occurrence of stone implements in the coast laterite south of Madras and in the higher level gravel and other formations in the South Mahratta country (abst.): *Brit. Ass. Rep.*, 1880, 589—590.

23. Notes on a traverse across the gold fields of Mysore *Rec. Geol. Surv. Ind.*, XV, 191—202, (1882).

Foote, R. B.,—cont.

24. On the Geology of the Madura and Tinnevelly districts : *Mem. Geol. Surv. Ind.,* XX, 1—103, (1883).

25. On the Geology of S. Travancore : *Rec. Geol. Surv. Ind.,* XVI, 20—35, (1883).

26. Rough notes on Billa Surgam and other caves in the Kurnool district ; *Rec. Geol. Surv. Ind.,* XVII, 27—34, (1884).

27. Mr. H. B. Foote's work at the Billa Surgam caves : *Rec. Geol. Surv. Ind.,* XVII, 200—208, (1884).

28. Notes on the country between the Singareni coal-field and the Kistna River : *Rec. Geol. Surv. Ind.,* XVIII, 12—25, (1885).

29. Geological sketch of the country between the Singareni coal-field and Hyderabad : *Rec. Geol. Surv. Ind.,* XVIII, 25—30, (1885).

30. Notes on the results of Mr. H. B. Foote's further excavations in the Billa Surgam caves : *Rec. Geol. Surv. Ind.,* XVIII, 227—235, (1885).

31. Notes on the geology of parts of Bellary and Anantapur districts : *Rec. Geol. Surv. Ind.,* XIX, 97, (1886).

32. Notes on some recent Neolithic and Palæolithic Finds in South India : *Jour. As. Soc. Beng.,* LVI, pt. ii, 259—282, (1887).

33. Report on the auriferous tracts in Mysore. Bangalore, 1887.

Foote, R. B., *and* **King, W.**

1. On the geological structure of parts of the districts of Salem, Trichino-poly, Tanjore, and South Arcot, in the Madras Presidency (being the area included in sheet 79 of the Indian Atlas) : *Mem. Geol. Surv. Ind.,* IV, 223—386, (1864).

Forbes, D.

1. Report on certain samples of iron ore from the Chanda district, Central Provinces of India : *Ind. Economist,* III, 78, (1871) ; *Gazette of India Supplement,* 1871, p. 1342.

Forbes, Edward.

1. Report on the collection of (Cretaceous) fossils from Southern India, presented by C. J. Kaye, Esq. and Rev. W. H. Egerton : *Quart. Jour. Geol. Soc.,* I, 79—81, (1845) ; *Cal. Jour. Nat. Hist.,* VI, 263—266, (1846).

2. Report on Fossil Invertebrata from Southern India, collected by Mr. Kaye and Mr. Cunliffe : *Geol. Trans.,* 2nd series, VII, 97—174, (1846).

Forbes, L. R.

1. Report on the Ryotwaree settlement of the Government Farms in Palamow, 8°. Calcutta, 1872. [Coal and Copper.]

Forlong, J. G. *and* **Fraser, A.**

1. Report on a Route from the mouth of the Pakchan to Kraw, and thence across the Isthmus of Kraw to the Gulf of Siam : *Jour. As. Soc. Beng.,* XXXI, 347—362, (1862) ; INDO-CHINA, No. 1, I, 285—297, (1886).

Forrest, R. E.

1. Memorandum on improvements in the irrigation of the Deyrah Doon and Remarks on the drainage of the Eastern portion of the valley : *Prof. Pap. Ind. Eng.,* 1st series, I, 57—74, (1863).

Forsyth, *Sir* T. D.

1. Report of a mission to Yarkand in 1873, with historical and geographical information regarding the possessions of the Ameer of Yarkand. 4°. Calcutta, 1875.

Franklin, J.

1. Memoir on Bundelkund: *Trans. Roy. As. Soc.*, I, 259—281, (1826). [Mines and Minerals, p. 277.]

2. On the Diamond mines of Panna in Bundelkhand; *As. Res.*, XVIII, pt. i, 101—122, (1829); *Edin. Jour. Sci.*, V, 150—166, (1831).

3. On the geology of a portion of Bundelkhand, Baghelkhand, and the Districts of Sagar and Jebelpur: *As. Res.*, XVIII, 23—46, (1833); *Glean. Sci.*, I, 213—215, (1829).

4. On the coal-field of Palamú: *Glean. Sci.*, II, 217—220, (1830).

5. On the Geology of Bundelkhand, Baghelkund and Districts of Saugor and Jubbulpore: *Geol. Trans.*, 2nd series, III, 191—200, (1835); *Proc. Geol. Soc.*, I, 82—85, (1834).

Fraser, H.

1. Further particulars regarding the Dandapur meteorite: *Proc. As. Soc. Beng.*, 1878, p. 190.

Fraser, J. B.

1. Account of a journey to the sources of the Jumna and Bhagirathi rivers: *As. Res.*, XIII, 171—249, (1820).

2. Journal of a Tour through part of the Snowy Range of the Himalaya Mountains and to the sources of the Rivers Jumna and Ganges. 4°. London, 1820.

3. Notes to accompany a set of specimens from the Himalaya Mountains: *Geol. Trans.*, V, 60—72, (1820).

4. Description accompanying a collection of specimens made on a journey from Delhi to Bombay; *Geol. Trans.*, 2nd series, I, 141—161, (1824).

5. Notes made in the course of a voyage from Bombay to Bushire in the Persian Gulf; transmitted with a series of illustrative specimens: *Geol. Trans.*, 2nd series, I, 409—412, (1824).

Frenzel, A.

1. Mineralogisches aus dem Ostindischen Archipel: *Jahrb. K.-K. Geol. Reichs.*, Bd. XXVII, *Min. Mitth.*, Heft. iii, pp. 297—309, (1877); *Boll. Comp. Map. Geol. Españ.*, VI, 87—90, (1879).

Frere, *Sir* H. Bartle, E.

1. On the Geology of a part of Sind: *Quart. Jour. Geol. Soc.*, IX, 349—351, (1853); Western India, 530—532, (1857).

2. On the Runn of Cutch and the countries between Rajpootana and Sind: *Brit. As. Rep.*, XXXIX, pt. ii, p. 163, (1869).

3. Notes on the Runn of Cutch and neighbouring regions: *Jour. Roy. Geog. Soc.*, XL, 181—207, (1870).

Fryar, Mark.

1. A letter to the Proprietors and Managers of the coal mines in India. 12°. London, 1869.

Fryar, Mark,—cont.

2. Note on the iron ores at Goonjwai, Lohara, and Dewulgaon in the Chanda district: *Gazette of India Supplement,* 1871, p. 1341; *Ind. Economist,* III, 78, (1871).

3. Report on coal exploration in the Chanda district: *Ind. Economist,* III, 136-137, (1871).

4. Report on some mineraliferous localities of Tenasserim: *Ind. Economist,* IV, 42—43, (1872); BURMA, pp. 413—444, (1882).

5. [Coal at Moulmein]: *Ind. Economist,* IV, 130, (1872); BURMA, pp. 444, 460—462, (1882).

6. [Correspondence regarding Tenasserim minerals]: BURMA, pp. 445—449.

7. Report on minerals in the Amherst district of the Tenasserim division: BURMA, pp. 450—459.

8. Report on minerals in Shwegyeen, Toungoo and Iahpoon districts, Tenasserim division: BURMA, pp. 462—475, (1882); *Colliery Guardian,* XXX, 390, (1873).

Fryer, G. E.

1. [On Burmese Celts]: *Proc. As. Soc. Beng.,* 1872, p. 46.

Fulljames, G.

1. Recent discovery of fossil bones in Perim Island on the Cambay Gulf: *Jour. As. Soc. Beng.,* V, 289-290, (1836); *Bibl. Univ.,* IX, 198-199, (1837).

2. Section of the strata passed through in an experimental boring at the town of Gogah, in the Gujerat Peninsula, Gulf of Cambay: *Jour. As. Soc. Beng.,* VI, 786—788, (1837).

3. A visit in December, 1832, to the cornelian mines situated in the Rajpeepla Hills, eastward of Broach: *Trans. Bo. Geog. Soc.,* II, 74—78, (1835).

4. Note on the discovery of fossil bones of Mammalia in Kattiwar: *Jour. Bo. As. Soc.,* I, 30—43, (1841).

5. Observations on the Runn [of Cutch]: *Trans. Bo. Geog. Soc.,* VII, 127, (1846).

6. Remarks on a singular hollow, 12 miles in length, called Boke, situated in the Purantij Parganah of the Ahmedabad collectorate: *Trans. Bo. Geog. Soc.,* VII, 164—167, (1846).

7. A description of the salt-water lake called the Null, situated on the Isthmus of Kattywar: *Jour. Bo. As. Soc.,* V, 109—117, (1857).

8. Discovery of nummulitic limestone in the Rajpipla hills: *Jour. Bo. As. Soc.,* V, 624—626, (1857).

9. Geology of the North Bank of the Narbadda from Baroda Eastwards: *Jour. Bo. As. Soc.,* VI, 163-164, (1861).

Fulljames, G. *and* **Hugel,** *Baron* **Karl von.**

1. Recent discovery of fossil bones in Perim Island in the Cambay Gulf: *Jour. As. Soc. Beng.,* V, 288—291, (1836).

E   49

G

Gell, F.
1. The Hill Forts of the Deccan : *Good Words*, 1878, pp. 24—29.

Geoffroy Saint-Hilaire, *see* **Saint-Hilaire.**

Gerard, A.
1. Journal of a journey through the Himalayah mountains from Shipke to the frontiers of Chinese Tartary : *Edin. Jour. Sci.*, I, 41—52, 215—225, (1824).
2. Narrative of a journey from Soobathoo to Shipke in Chinese Tartary : *Jour. As. Soc. Beng.*, XI, 363—391, (1842).

Gerard, A. *and* **P.**
1. Account of a journey through the Himalaya Mountains : *Edin. New Phil. Jour.*, X, 295—305, (1824).

Gerard, A. *and* **Colebrooke, H. T.**
1. On the Valley of the Setlej River, in the Himalaya Mountains, from the Journal of Captain A. Gerard, with remarks by Henry Thomas Colebrooke. [A few scattered geological observations] : *Trans. Roy. As. Soc.*, I, 343—380, (1826) ; *Edin. Jour. Sci.*, V, 270—278 ; VI, 28—50, (1826).

Gerard, Alex. *and* **Lloyd, Sir W.**
1. Narrative of journey from Caunpore to the Boorendo Pass in the Himalaya Mountains, *via* Gwalior, Agra, Delhi, and Sirhind, by Major Sir William Lloyd : and an account of an attempt to penetrate by Bekhur to Garoo and the Lake Manasarowara, by Captain Alexander Gerard : with a letter from the late J. G. Gerard, detailing a visit to the Shaitool and Boorendo Passes for purpose of determining the line of perpetual snow on the southern face of the Himalayas, &c.   Edited by George Lloyd.   2 vols., 8°.   London, 1840.

Gerard, J. G.
1. Letter from the Himalayas. [Fossils] : *Glean. Sci.*, I, 92, 109-11, (1829).
2. Notice of the discovery of fossils in Thibet, 17,000 ft. above the sea, by J. G. Gerard : *Edin. New Phil. Jour.*, X, 178, (1831).
3. Observations on the Spiti valley and the circumjacent country within the Himalaya : *As. Res.*, XVIII, 238—278, (1832).
4. Fossil Shells near Herat : *Jour. As. Soc. Beng.*, II, 652, (1833).

Gerard, J. G. *and* **Burnes, Alex.**
1. Sketch of the Route and Progress of Lieut. A. Burnes and Dr. Gerard : *Jour. As. Soc. Beng.*, I, 139—145, (1832) ; II, 1—22, 143—149, (1833).

Gerard, P.
1. Remarks on some of the mineral productions of the Himalayas : *Delhi Medical Journal*, I, 62-71, (1844).

Gibson, A.

1. A general sketch of the Province of Guzerat, from Deesa to Damaun: *Trans. Med. Phys. Soc. Bombay,* I, 1—77, (1838).

2. Report on the iron ore found at Malwan: *Jour. Bo. As. Soc.,* I, 142—144, (1842).

Gibson, J.

1. On the composition of "Reh," an efflorescence on the soil of certain districts of India: *Proc. Roy. Soc. Edin.,* X, 277—290, (1879).

Gilchrist, P. C. *and* Riley, E.

1. The iron-making Resources of the British Colonies as illustrated at the Colonial and Indian Exhibition; India: *Iron,* XXVIII, 476—478, (1886).

Gilchrist, W.

1. On the origin and formation of the Red soil of Southern India: *Quart. Jour. Geol. Soc.,* XI, 552—555, (1855).

Giraud, H.

1. A chemical and microscopical examination of the Rock salt of the Punjab: *Jour. Bo. As. Soc.,* I, 303—308, (1843).

2. An account of two aërolites, and a mass of Meteoric Iron, recently found in Western India: *Edin. New Phil. Jour.,* XLVII, 53—57, (1849); *Froriep. Notizen,* 241—244, (1849).

Giraud, H. *and* Haines, R.

1. Analysis of the Mineral Springs and various Well Waters in the Bombay Presidency: *Trans. Med. Phys. Soc. Bombay,* V, 242—263, (1859).

Glasfurd, J.

1. Report on the progress made up to the 1st May, 1839, in opening the experimental Copper Mine in Kumaon: *Jour. As. Soc. Beng.,* VIII, 471—474, (1839).

Glass.

1. Correspondence and papers relative to the manufacture of glass and Earthenwares and Fire-bricks in India: *Cal. Jour. Nat. Hist.,* II, 589—609, (1842).

2. List of some of the articles which have been made at the Fattehpur Flint Ware Factory: *Cal. Jour. Nat. Hist.,* III, 152, (1843).

Gmelin, C. G.

1. Analysis of cinnamon stone from Ceylon: *Edin. Phil. Jour.,* XI, 129—132, (1824).

2. Analysis of a black mineral from Candy in Ceylon, named Candite: *Edin. Phil. Jour.,* IX, 384—387, 1823; *Ann. de Chemie,* XXV, 208—209, (1824).

3. Chemische untersuchung des Poonahlits und des Thulits: *Annalen, Phys. Chim.,* 2nd series, XLIX, 538—540, (1840).

Godwin Austen, H. H., *see* AUSTEN, H. H. GODWIN.

Gold.

1. Gold Mines at Malacca. Extract from the Proceedings of the Coal Committee: *Cal. Jour. Nat. Hist.,* IV, 539, (1844).

Gold,—cont.

    2. Account of the gold mines in the Province of Malabar, from official papers communicated by Government : *Mad. Jour. Lit. Sci.*, XIV, 154—181, (1847).

    3. Correspondence regarding gold mines in Wynaad, Malabar district. 8° Madras, 1874.

    4. The gold quartz regions of Southern India : *Mining Journal*, L, 67—68, (1880).

Goldsmid, *Sir* **F. J.**

    1. Eastern Persia, an account of the journeys of the Persian Boundary Commission, 1870-71-72. 2 vols. 8vo. London, 1876. [Vol. II, Zoology and Geology by W. T. Blanford].

Gordon, A.

    1. Notes on the Topography of Murree : *Jour As. Soc. Beng.*, XXIII, 461—469, (1854).

Gordon, R.

    1. Report on the Irrawaddy river. Folio. Rangoon, 1879—1880.

Grange, E. R.

    1. Extracts from the Narrative of an Expedition into the Naga territory of Assam : *Jour. As. Soc. Beng.*, VIII, 445—470, (1839) ; IX, 947—966, (1840).

Grange, Jules.

    1. Géologie, Minéralogie, et Géographie Physique du Voyage, au Pole Sud et dans l' Océanie sur les corvettes *l' Astrolabe* et *la Zélée*, exécuté pendant les années 1837—1840, d'après les matériaux recueillis par MM. les chirurgiens naturalistes d'expédition. 8°. Paris, 1847—1848, with folio atlas.

Grant, C. W.

    1. Progress of the Boring for coal at Jamutra in Cutch : *Jour. As. Soc. Beng.*, III, 40—42, (1834).

    2. Memoir to illustrate a geological map of Cutch : *Geol. Trans.*, 2nd series, V, 289—329, (1840) ; *Mad. Jour. Lit. Sci.*, XII, 307—374, (1840) ; Western India, 403—459, (1857).

Grant, F. T.

    1. Mode of extracting the Gold Dust from the sands of the Ningthee river : *Jour. As. Soc. Beng.*, I, 148—149, (1832).

    2. Extracts from a Journal kept during a Tour of Inspection of the Manipúr Frontier, along the course of the Ningthee River, &c., in January 1832 : *Jour. As. Soc. Beng.*, III, 124—134, (1834).

Gray, J. E.

    1. Notice of a fossil Hydraspide (*Testudo Leithii*, Carter) from Bombay : *Ann. Mag. Nat. Hist.*, 2nd series, VIII, 339—340, (1871).

Gray, O. W.

    1. On the discovery of fossil Bones near Hingoli [Godavery valley] : *Mad. Jour. Lit. Sci.*, VII, 477, (1838).

Greenough, G. B.

1. General sketch of the physical and geological features of British India: London, (1854). [Geological map.]

2. On the geology of India: *Bril. Ass. Rep.*, 1854, pt. ii, pp. 83-85 ; *Ann. des Mines*, VI, 577—586, (1854) ; *Peterman, Mittheil.* 1885, pp. 23—27.

Greenough's Map of India.

1. [Report of a committee appointed to consider Prof. Greenough's Geological Map of India]: *Jour. As. Soc. Beng.*, XXV, 419—426, (1856).

2. Correspondence on the subject of the geological map of India, compiled by Prof. Greenough. 8°. Madras, 1857.

Greenwood, G.

1. Note on Mr. Drew's account of the alluvial deposits of the Upper Indus basin: *Geol. Mag.*, 2nd decade, I, 45, (1874).

Gregory, W.

1. On the composition of Petroleum of Rangoon, with remarks on Petroleum and Naphtha in general: *Trans. Roy. Soc. Edin.*, XIII, 124—140, (1836); *Jour. As. Soc. Beng.*, IV, 527—528, (1835); *Jour. de Pharm.*, XXI, 536-541, (1836).

Griesbach, C. L.

1. Geology of the Ramkola and Tatapani coalfields: *Mem. Geol. Surv. Ind.*, XV, 129-192, (1880).

2. Geological notes [Himalayas north of Kumaon]: *Rec. Geol. Surv. Ind.*, XIII, 83—93, (1880).

3. Palæontological notes on the Lower Trias of the Himalayas: *Rec. Geol. Surv. Ind.*, XIII, 94—113, (1880); XIV, 154—155, (1881).

4. On the Geology of the section between the Bolan Pass in Biluchistan and Girishk in Southern Afghanistan: *Mem. Geol. Surv. Ind.*, XVIII, 1—60, (1881).

5. Report on the Geology of the Takht-i-Suliman: *Rec. Geol. Surv. Ind.*, XVII, 175—190, (1884).

6. Afghan field notes: *Rec. Geol. Surv. Ind.*, XVIII, 57—64, (1885).

7. Afghan and Persian Field notes: *Rec. Geol. Surv. Ind.*, XIX, 48—65, (1886).

8. Field notes from Afghanistan, (No. 3) Turkistan: *Rec. Geol. Surv. Ind.*, XIX, 235—268, (1886).

9. Field notes from Afghanistan, (No. 4); from Turkistan to India: *Rec. Geol. Surv. Ind.*, XX, 17—26, (1887).

10. Field notes, No. 5 ; to accompany a Geological Sketch map of Afghanistan and North-Eastern Khorasan: *Rec. Geol. Surv. Ind.*, XX, 93—103, (1887).

11. Notice of J. B. Mushketoff's Geology of Russian Turkistan. Compiled from translation and notes of Prof. F. Toula of Vienna: *Rec. Geol. Surv. Ind.*, XX, 123—134, (1887).

Griffith, W.

1. Journal of a visit to the Mishmee Hills in Assam: *Jour. As. Soc. Beng.,* VI, 325—341, (1837).

2. Journal of a mission which visited Bootan, in 1837—38, under Capt. R· Boileau Pemberton; *Jour. As. Soc. Beng.,* VIII, 208—241, 251—291' (1839).

3. Extracts from a report on subjects connected with Afghanistan; *Jour. As. Soc. Beng.,* X, 797—815, 977—1037, (1841).

4. Journals of Travels in Assam, Burma, Bootan, Afghanistan, and the neighbouring countries: posthumous papers, arranged by Dr. J. McClelland, 8° Calcutta, 1847.

Gubbins, C.

1. Mode of manufacture of the Salumba salt of Upper India: *Jour. As. Soc. Beng.,* VII, 363—364, (1838).

Gumbel, C. W.

1. Ueber das Vorkommen von unteren Trias schicten in Hochasien, [Ammonites collected by the Schlagintweits reputed to be from Dharampur]: *Sitz. Bair. Accad. Wiss.* 1865, Band II, pp. 348—366.

Gunther, A.

1. Note on a Fish plate from the Sivaliks: *Rec. Geol. Surv. Ind.,* XIV, 240, (1881).

Guyerdet, A.

1. Déterminations de quelques coquilles fossiles du Thibet. [Devonian fossils sent by M. Thomine des Mazures from Eastern Thibet]: *Comptes Rendus,* LVIII, 879, (1864).

H

Hacket, C. A.

1. Geology of Gwalior and its vicinity: *Rec. Geol. Surv. Ind.,* III, 32—42, (1870).

2. Note on the Arvali series in North-Eastern Rajputana: *Rec. Geol. Surv. Ind.,* X, 84—92, (1877).

3. Salt in Rajputana: *Rec. Geol. Surv. Ind.,* XIII, 197—206, (1880).

4. Useful minerals of the Aravali region: *Rec. Geol. Surv. Ind.,* XIII, 243—250, (1880).

5. On the Geology of the Aravali region, Central and Eastern: *Rec. Geol. Surv. Ind.,* XIV, 279—303, (1881).

Hackney, W.

1. Report on the Iron Ore and Coal from the Chanda District of the Central Provinces of India: *Prof. Pap. Ind. Eng.,* 2nd series, IX, 185—221, (1880).

Haidinger, W.

1. [Notes on some Indian meteorites; Piddingtonite in the Shalka Meteorite]: *Jour. As. Soc. Beng.,* XXIX, 416—418, (1860).

Haidinger, W.,—cont.

2. Der Meteorit von Shalka in Bancoorah, und der Piddingtonit : *Sitz. K. K. Akad. Wien*, XLI, 251—260, (1860).

3. Die Calcutta—Meteoriten von Shalka, Futtehpore, Pegu, Assam und Segowlee im K. K. Hof Mineralien cabinete : *Sitz. K. K. Akad. Wien*, XLI, 745—758, (1860) ; *Jour. As. Soc. Beng.*, XXX, 129—138, (1861).

4. Die Meteoritenfälle von Quengouk bei Bassein in Pegu und Dhurmsala im Punjab : *Sitz. K. K. Akad. Wien*, XLII, 301—306, (1860).

5. Das Doppelmeteor von Elmira und Long Island : *Sitz. K. K. Akad. Wien*, XLIII, abth. 2, 304—307, (1861).

6. Der Meteorit von Yatoor bei Nellore in Hindoostan : *Sitz. K. K. Akad. Wien*, XLIV, abth. 2, 73—74 (1861).

7. Der Metorit von Parnallee bei Madura im K. K. Hofmineralien cabinete : *Sitz. K. K. Akad. Wien*, XLIV, abth. 2, 117—120, (1861).

8. Der Meteorit von Dhurmsala : *Sitz. K. K. Akad. Wien*, XLIV, abth. 2, 285—288, (1861).

9. Das Meteor von Quenggouk in Pegu und die Ergebnise des Falles darselbst am 27 Decembre 1857 : *Sitz. K. K. Akad. Wien*, XLIV, abth. 2, 637—642, (1862).

10. Der Meteorsteinfall in Goruckpur Districte in Ober Bengalen am 12 Mai 1861 : *Sitz. K. K. Akad. Wien*, XLV, abth. 2, 665—671, (1862).

11. Das Eisen von Kurruckpur nicht meteorischen Ursprungs : *Sitz. K. K Akad. Wien*, XLVIII, abth. 2, 672—674, (1863).

12. Ueber den Fall von Meteorstein bei Parnallee in Ost Indien *Sitz. K. K. Akad. Wien*, XLVII, abth. 2, 420—426, (1863).

13. Der Meteorstein von Manbhoom in Bengalen im K. K. Hof Mineralien Cabinete aus dem Falle am 22 Decembre 1863 : *Sitz. K. K. Akad. Wien*, L, abth. 2, 241—246, (1865).

14. Hessle, Rutlam, Assam, drei neue Meteoriten : *Sitz. K. K. Akad. Wien*, LIX, abth. 2, 224—230, (1869).

15. Der Meteorit von Goalpara in Assam nebst Bemerkungen über die Rotation der Meteoriten in ihrem Zuge : *Sitz. K. K. Akad. Wien*, LIX, abth. 2, 665—678, (1869).

Haines, S. B.

1. A description of the Arabian Coast, commencing from the entrance of the Red Sea and continuing as far as Massenat : *Trans. Bo. Geog. Soc.*, XI, 60—276, (1854).

Hale, A.

1. On mines and miners in Kinta, Perak : *Jour. Straits As. Soc.*, 1886, pp. 303—320.

Hallet, Holt S.

1. Exploration Survey for a railway between India, Siam, and China : *Scottish Geog. Mag.*, II, 78—91, (1886).

Halstead, E. P.

1. Report on the Island of Chedooba : *Jour. As. Soc. Beng.*, X, 349—376 419—435, (1841).

55

Hamilton, C.

A short description of the Car Nicobar : *As. Res.*, II, 337—344, (1799).

Hamilton, Francis, *see* BUCHANAN.

Hamilton, *Sir* R. N. C.

1. Note on the Transport of coal from the pits at Sonadeh to Bombay by the Nerbudda : *Jour. As. Soc. Beng.*, XVIII, 594—600, (1849).

2. Report on a trip down the Nerbudda from Mundlaisir to Baroche : *Trans. Bo. Geog. Soc.*, VIII, 119—143, (1849).

3. Table of heights and distances along the proposed Railway from Surat to Agra : *Jour. As. Soc. Beng.*, XXV, 221, (1856).

Hamilton, W.

1. A Geographical, Statistical, and Historical description of Hindustan, and the adjacent countries. 2 vols., 4°. London, 1820.

Hampton, J. H.

1. Cassiterite of the Straits Settlements : *Mineral Mag.*, VII, 71, (1886).

Hannay, S. F.

1. Announcement of two new sites of Coal in Assam : *Jour. As. Soc. Beng.*, VII, 169, (1838).

2. Note respecting the Jaipur [Assam] coal : *Jour. As. Soc. Beng.*, VII, 368, (1838).

3. Further information on the Gold Washings of Assam : *Jour. As. Soc. Beng.*, VII, 625—628, (1838).

4. Memoranda of earthquakes and other remarkable occurrences in Upper Assam, from January 1839, to September 1843 : *Jour. As. Soc. Beng.*, XII, 907—908, (1843).

5. On the Assam Petroleum beds : *Jour. As. Soc. Beng.*, XIV, 817—820, (1845).

6. Extracts from letters respecting the Jaipur [Assam] coal : *Jour. As. Soc. Beng.*, XVII, 167—168, (1848).

7. Notes on the Gold fields of Assam : *Jour. As. Soc. Beng.*, XXII, 515—521, (1853).

8. Notes on the productive capacities of the Shan countries, North and East of Ava : *Sel. Rec. Beng. Gov.*, XXV, 9—19, (1857).

9. Notes on the Iron ore statistics, and Economic Geology of Upper Assam : *Jour. As. Soc. Beng.*, XXV, 330—344, (1856).

Hannay, S. F. *and* Dalton, E. T.

1. Note on recent investigations regarding the extent and value of the Auriferous deposits of Assam, being abstracts of reports by Captain E.T. Dalton and Lieutenant-Colonel S. F. Hannay, dated October, 1855 : *Mem. Geol. Sur. Ind.*, I, pt. i, 94—98, (1856).

Hannay, S. F. *and* Pemberton, R. B.

1. Abstract of the Journal of a Route, travelled by Captain S. F. Hannay of the 48th Regiment, Native Infantry, from the Capital of Ava to the Amber Mines of the Húkong valley on the South-East Frontie of Assam : *Jour. As. Soc. Beng.*, VI, 245—278, (1837).

Harcourt, A. F. P.

1. The Himalayan Districts of Kooloo, Lahoul, and Spiti: *Sel. Rec. Punjab Gov.*, new series, X, 79—80, (1874); *also* 8°. London, 1871.

Hardie, J.

1. Observations on the Geology of the Meywar District: *Edin. New Phil. Jour.*, VI, 329—336; VII, 116—125, (1829).

2. On the Geology of the Secondary Formations of the Meywar District: *Edin. New Phil. Jour.*, XI, 82—90, (1831).

3. Outline of the Geology of the Bhurtpore District: *Edin. New Phil. Jour.*, XIII, 328—337, (1832); XIV, 76—82, (1833).

4. Geology of the valley of Oodipur: *Edin. New Phil. Jour.*, XIV, 263—283, (1833); XVI, 56—67, 278—285, (1834).

5. Remarks on the Geology of the country on the Route from Baroda to Udayapur, *vid* Birpur and Salambhar: *As. Res.*, XVIII, pt. i, 82—99, (1833); *Glean. Sci.*, I, 218—220, (1829).

6. Sketch of the Geology of Central India, exclusive of Malwa: *As. Res.*, XVIII, pt. ii, 27—92, (1833).

7. Explanation of the sketch giving a Geological section of the Strata from Nimach to Merta, published in the *Asiatic Researches*, Vol. XVIII, pt. ii, p. 92: *Jour. As. Soc. Beng.*, III, 238—242, (1834).

Harris, J. C.

1. Notes on the Rainfall in the Basin of the River Mahanuddy and the floods consequent thereupon: *Jour. As. Soc. Beng.*, XXX, 216—220, (1861).

Hart, N.

Some account of a Journey from Kurrachee to Hinglaj, in the Lus territory, descriptive of the intermediate country and of the port of Soumeanee: *Jour. As. Soc. Beng.*, IX, 134—154, 615, (1840).

Haughton, J. C.

1. Memorandum on the Geological Structure and Mineral Resources of the Singhbhoom Division, South-West Frontier Agency: *Jour. As. Soc. Beng.*, XXIII, 103—122, (1854).

2. Account of a Meteor in Cooch Behar, 30th April, 1869: *Proc. As. Soc. Beng.*, 1866, p. 169.

Haughton, S.

1. On Hypostilbite and Stilbite [Stilbite from Nerbudda]: *Phil. Mag.*, 4th ser., XIII, 509—510, (1857).

2. Mineralogical description of rocks collected near Nagpur, Central India, by the Rev. Messrs. Hislop and Hunter: *Jour. Roy. Dub. Soc.*, II, 175—180, (1858-1859); *Nat. Hist. Rev.*, VI, (Proc.) 42—47, (1859).

3. On some Rocks and Minerals from Central India, including two new species, Hislopite and Hunterite: *Phil. Mag.*, XVII, 4th ser., 16—21, (1859).

4. Additional notice of Hislopite and Hunterite: *Phil. Mag.*, XXIII, 50—51, (1862).

Haughton, S.,—cont.

5. On the chemical composition of four Zeolites, presented by Col. Montgomery to the Geological Museum of Trinity College, Dublin : *Phil. Mag.*, 4th ser., XXXII, 223—225, (1866); *Jour. Roy. Geol. Soc. Dublin*, I, 252—254, (1867).

6. On the Chemical and Mineralogical composition of the Dhurmsala meteoric stone : *Proc. Roy. Soc.*, XV, 214—218, (1867); *Phil. Mag.*, XXXII, 266—269, (1866).

7. Additional notice of the Zeolites of Western India : *Jour. Roy. Geol. Soc. Dublin*, II, (new series), 112—115, (1871).

8. On the chemical composition of some Zeolites from Bombay : *Jour. Roy. Geol. Soc. Dublin*, I, (new series), 252—256,(1868).

Hay, R. G.

1. Fossil shells discovered by Capt. Hay in the neighbourhood of Bajgah, Afghanistan : *Jour. As. Soc. Beng.*, IX, 1126—1128, (1840).

2. Report on the valley of Spiti ; and facts collected with a view to a future Revenue Settlement : *Jour. As. Soc. Beng.*, XIX, 429—448, (1850).

Hay, R. G. *and* Thurburn, F. A. V.

1. Report on the Turan Mull Hill [Satpuras] : *Jour. As. Soc. Beng.*, XX, 502—516, (1851).

Heath, J. M.

1. [Memoranda on the subject of the Salem Iron works] : *Jour. As. Soc. Beng.*, I, 253—255, (1832).

2. On Indian Iron and Steel : *Jour. Roy. As. Soc.*, V, 390—397, (1839); *Mad. Jour. Lit. Sci.*, XI, 184—191, (1840); *Appendix Rep. Gov. Mus.*, *Madras*, 1856, pp. 1—8.

Heatley, S. G. I.

1. Contribution towards a History of the Development of the Mineral Resources of India : *Jour. As. Soc. Beng.*, XI, 811—835, (1842).

2. Contribution, &c., No. 2. Memoranda relative to the working of Iron in Bengal : *Jour. As. Soc. Beng.*, XII, 542—545, (1843).

Heber, *Bishop* R.

1. Narrative of a Journey through the Upper Provinces of India from Calcutta to Bombay, 1824—1825 (with notes on Ceylon) ; an account of a Journey to Madras and the Southern Provinces, 1826 ; and Letters written in India. With plates. 2 vols., 4°. London, 1828.

Heddle, J. F.

1. Memoir on the River Indus : *Sel. Rec. Bo. Gov.*, XVII, 401—457, (1855).

Helfer, J. W.

1. [Letter on tin, iron, &c., from Tenasserim] : *Jour. As. Soc. Beng.*, VII, 171, (1838).

2. Report on the Coal discovered in the Tenasserim Provinces : *Jour. As. Soc. Beng.*, VII, 701—706, (1838).

Helfer, J. W.,—cont.

3. Second report on the Provinces of Ye, Tavoy, and Mergui, on the Tenasserim Coast, visited and examined by order of Government, with the view to develope their natural resources. 8°. Calcutta, 1839, (Reprint, 1875.)

4. Report on the coal-field at Tha-thay-yna on the Tenasserim River, in Mergui Province : *Jour. As. Soc. Beng.*, VIII, 385—389, (1839).

5. Gedruckte und ungedruckte Schriften über die Tenasserim Provinzen, den Mergui Archipel und die Andamanen Inseln : *Mittheil. K. K. Geog. Gesel.*, III, 166—390, (1859).

Helfer, Pauline.

1. Travels of Dr. and Madame Helfer in Syria, Mesopotamia, Burma and other lands. Rendered into English by Mr. George Sturge. 2 vols., 8°. London, 1878.

Henderson, G.

1. Notes on Sand Pits, Mud Discharges, and Brine Pits met with during the Yarkand Expedition of 1870 : *Quart. Jour. Geol. Soc.*, XXVIII, 402—404, (1872).

Henderson, G. *and* Hume, A. O.

1. Lahore to Yarkand. Incidents of the route and natural history of the countries traversed by the Expedition of 1870 under T. D. Forsyth, Esq., C.B. 8°. London, 1873.

Henderson, W.

1. Memorandum on the nature and effects of the flooding of the Indus on the 10th August, 1858, as ascertained at Attock and its neighbourhood : *Jour. As. Soc. Beng.*, XXVIII, 199—219, (1859).

Henwood, W. J.

1. Report on the metalliferous deposits of Kumaon and Garhwal : *Sel. Rec. Gov. Ind.*, VIII, 1—46, (1855).

2. On the metalliferous deposits of Kumaon and Garhwal in North-Western India : *Edin. New Phil. Jour.*, 2nd series, III, 135—141, (1856).

Herbert, J. D.

1. An account of a Tour made to lay down the Course and Levels of the River Sutlej or Satúdra as far as traceable within the limits of British Authority, performed in 1819 : *As. Res.*, XV, 339—428, (1825).

2. On the Zehr Mohereh, or Snake Stone : *As. Res.*, XVI, 382—386, (1828)·

3. Notice of the occurrence of Coal within the Indo-Gangetic tract of mountains : *As. Res.*, XVI, 397—408, (1828).

4. Particulars of a visit to the Siccim Hills with some account of Darjiling, a place proposed as the site of a sanitarium, or Station of health : *Glean. Sci.*, II, 89—96 & 114—124, (1830).

5. On the accumulation of Diluvium or Gravel in the valleys which border the great Himalayan system of Formation [well-section in the Dehra-Dun] : *Glean. Sci.*, II, 164—165, (1830).

6. On the Organic Remains found in the Himalayas : *Glean. Sci.*, III 265—272, (1831).

Herbert, J. D.,—cont,

7. Notice of the occurrence of gypsum in the Indo-Gangetic tract of moun⁻ tains: *As. Res.*, XVIII, pt. i, 216—233, (1833); *Glean. Sci.*, I, 227—228, (1829).

8. On the Mineral Productions of that part of the Himálaya Mountains lying between the Sutlej and the Kali (Gàgra) Rivers, considered in an economical point of view ; including an account of the mines and methods of working them, with suggestions for their improvement : *As. Res.*, XVIII, pt. i, 227—258, (1833); *Glean. Sci.*, I, 228—230, (1829).

9. Report on the Mineralogical Survey of the Himalaya Mountains lying between the Rivers Sutlej and Kali. Illustrated by a Geological Map : *Jour. As. Soc. Beng.*, XI, Supplement pp. i—clxiii, (1842). Map XIII, pt. i, 171, (1844).

Herbert, J. D., *and* **Batten, J. H.**

1. Journal of Captain Herbert's Tour from Almorah in a N.W , W., and S.W. direction through parts of the province of Kumaon and British Gurhwal chiefly in the centre of the Hills, *vide* No. 66, Indian Atlas. Edited by J. H. Batten : *Jour. As. Soc. Beng.*, XIII, 734—764, (1844).

Heyne, B.

1. Tracts, historical and statistical, on India ; with journals of several tours through various parts of the Peninsula ; also an account of Sumatra, in a series of letters. 4°. London, 1814.

2. On the formation of sulphur in India : *Phil. Mag.*, XLI, 101—104, (1813).

Himalayas.

1. Account of a visit to the Bians pass in the Indo-Gangetic Range beyond the head of the Kali River : *Glean. Sci.*, I, 97—99, (1829).

2. [Supposed] Volcano in the Himalayas. [Nanda Devi]: *Glean. Sci.*, I, 338, (1829).

3. Letter from the Himalayas—[signed I. A. H.]: *Glean. Sci.*, II, 48—52, (1830).

Hinton.

1. [Coal at Ramree] : *Cal. Jour. Nat. Hist.*, II, 115—117, (1842).

Hislop, S.

1. On the age of the Coal Strata in Western Bengal and Central India : *Jour. As. Soc. Beng.*, XXIV, 347—353, (1855).

2. [Gondwana and Intertrappean fossils from Nagpur]: *Jour. As. Soc. Beng.*, XXIV, 365, (1855).

3. On the connection of the Umret coal beds with the Plant beds of Nagpur and of both with those of Burdwan : *Quart. Jour. Geol. Soc.*, XI, 380, 555—561, (1855) ; Western India, 291—300, (1857).

4. Geology of the Nagpur State : *Jour. Bo. As. Soc.*, V, 58—77, 148—150, (1857).

5. On the Tertiary Deposits associated with the Trap rock in the East Indies, with a description of the Fossil Shells : *Quart. Jour. Geol. Soc.*, XVI, 154—182, (1860).

6. Remarks on the Geology of Nagpur : *Jour. Bo. As. Soc.*, VI, 194—206, (1861).

Hislop, S.,—cont.

7. On the age of the Fossiliferous thin-bedded sandstone and Coal of the Province of Nagpur, India: *Quart. Jour. Geol. Soc.,* XVII, 346—354, (1861).

8. Supplementary note on the Plant bearing sandstones of Central India: *Quart. Jour. Geol. Soc.,* XVIII, 36, (1862).

9. Extracts from letters relating to the further discovery of fossil Teeth and Bones of Reptiles in Central India: *Quart. Jour. Geol. Soc.,* XX, 117, 280—282, (1864).

10. [Note on the geological position of the Bijori Labyrinthodont]: *Jour. As. Soc. Beng.,* XXXIII, 443, (1864).

Hislop, S. *and* **Hunter, R.**

1. On the Geology and Fossils of the neighbourhood of Nagpur, Central India: *Quart. Jour. Geol. Soc.,* X, 470, (1854); XI, 345—383, (1855); WESTERN INDIA, 247—287, (1857).

Hochstetter, F. Von.

1. Beiträge zur Geologie und physikalischen Geographic der Nikobar—Inseln: NOVARA, II, 83—112, (1864); *Rec. Geol. Surv. Ind.,* II, 59—73, (1869).

2. Asien, seine zukuftsbahnen und seine kohlenschätze; 8°. Vienna, 1876. *see* MOSA; *Sel. Rec. Gov. Ina.,* LXXVII, 208-230, (1870).

Hodges, J. F.

1. On the composition of Tea and Tea-Soils from Cachar: *Brit. Ass. Rep.,* 1874, pt. ii, pp. 60—63.

Hodgson, B. H.

1. On the Physical Geography of the Himalaya: *Jour. As. Soc. Beng.,* XVIII, 761—788, (1849); *Sel. Rec. Beng. Gov.,* XXVII, 48—82, (1857).

Hodgson, C. K.

1. Memorandum on earthquakes in Jan. 1849, at Burpetah, Assam: *Jour. As. Soc Beng.,* XVIII, 174—175, (1849).

Hodgson, J. A.

1. Journal of a survey to the Heads of the Rivers Ganges and Jumna: *As. Res.,* XIV, 60—152, (1822); *Ann. Phil.,* IV, 31—52, 99—117, (1822); *Edin. Phil. Jour.,* VIII, 231—244; IX, 7—20, (1823).

Hoff, W. H.

1. Précis of information regarding the Andamans, Nicobars and Coco Islands, prepared in the Foreign Office: *Sel. Rec. Gov. Ind.,* XXV, 53—71, (1859).

Hoffmeister, W.

1. Travels in Ceylon, and continental India; including Nepal and other parts of the Himalayas, to the borders of Thibet, with some notices of the overland route. 8°. Edinburgh, 1848.

Hogg, J.

1. On Gebel Haurâin, its adjacent districts, and the Eastern Desert of Syria, with remarks on their geography and geology: *Edin. New Phil. Jour.,* 2nd series, XI, 173—192, (1860).

Holdich, T. H.

1. Note on the Sambar Salt Lake: *Sel. Rec. Gov. Ind.*, LXXI, 147—149, (1869).

Homfray, J.

1. A description of the coal-field of the Damooda Valley and the adjacent countries of Beerbhoom and Pooroooliah, as applicable to the present date, 1842: *Jour. As. Soc. Beng.*, XI, 723—739, (1842).

2. [Ball coal of Burdwan]: *Jour. As. Soc. Beng.*, XVII, pt. ii, 703 (1848).

Hooker, *Sir* J. D.

1. Observations made when following the Grand Trunk Road across the hills of Upper Bengal, Parus Nath &c., in the Soane Valley; and on the Kymaon branch of the Vindhya Hills: *Jour. As. Soc. Beng.*, XVII, pt. ii, 355—411, (1848).

2. Notes, chiefly Botanical, made during an excursion from Darjiling to Tonglo, a lofty mountain on the confines of Sikkim and Nepal: *Jour. As. Soc. Beng.*, XVIII, 419—446, (1849).

3. On the physical character of the Sikkim Himalaya: *Lond. Jour. Bot.*, III, 21—31, (1851).

4. A fourth excursion to the passes into Tibet by the Donkiah Lah: *Jour. Roy. Geog. Soc.*, XX, 49—52, (1851).

5. Himalayan Journals; or notes of a naturalist in Bengal, the Sikkim and Nepal Himalayas, the Khasia Moutains, &c.   2 vols., 8°. London, 1854.

Host, H.

1. [Notice of garnet mines in Rajputana]: *Sel. Rec. Gov. Ind.*, LXXX, 70—74, (1870).

Hoste, E. P. de La, *see* DE LA HOSTE.

Hot Springs.

1. Temperature of the, at Peer Mungal or Munga or Mungear: *Jour. As. Soc. Beng.*, XVII, pt. ii, 230, (1848).

Hove.

1. Tours for Scientific and Economical Research made in Guzerat, Kattiawar, and the Conkuns: *Sel. Rec. Bo. Gov.*, new series, XVI, (1855).

Hugel, *Baron* Karl von.

1. Notice of visit to the Valley of Kashmir in 1836 : *Jour. As. Soc. Beng.*, V, 184—187, (1836).

2. Notice of a visit to the Himmaleh Mountains and the Valley of Kashmir in 1835: *Jour. Roy. Geog. Soc.*, VI, 343—350, (1836).

3. [Letter on Fossil bones from Perim] : *Jour. Bo. As. Soc.*, I, 17—18, (1841).

4. Kaschmir und das Reich der Siek.   4 vols., 8vo. Stuttgart, 1840—1848.

5. Das Kabul Becken und die Gebirge Zwischen dem Hindu Kosch und der Sutlej.   4°. Wien, 1850.

Hugel, *Baron* Karl von *and* Fulljames, G.

1. Recent discovery of Fossil bones in Perim Island in the Cambay Gulf: *Jour. As. Soc. Beng.*, V, 288—291, (1836).

Hughes, A. W.

1. The country of Baloochistan, its Geography, Topography, Ethnology and History. 8°. London, 1877.

Hughes, T. W. H.

1. The Jherria coal-field : *Mem. Geol. Surv. Ind.*, V, 227—336, (1866).
2. The Bokaro coal-field : *Mem. Geol. Surv. Ind.*, VI, 39—108, (1867).
3. The Kurhurbari coal-field : *Mem. Geol. Surv. Ind.*, VII, 209—246, (1870).
4. The Deoghur coal-fields : *Mem. Geol. Surv. Ind.*, VII, 247—255, (1870).
5. Note on Slates at Chitéli, Kumaon : *Rec. Geol. Surv. Ind.*, III, 43—44, (1870).
6. On the lead ore at Slimanabád, Jabalpur District, Central Provinces: *Rec. Geol. Surv. Ind.*, III, 70—71, (1870).
7. The Karanpura coal-fields : *Mem. Geol. Surv. Ind.*, VII, 285—342, (1871).
8. The Itkhuri coal-field : *Mem. Geol. Surv. Ind.*, VIII, 321—324, (1872).
9. The Daltonganj coal-field : *Mem. Geol. Surv. Ind.*, VIII, 325—346, (1872).
10. Coal in India : *Rec. Geol. Sur. Ind.*, VI, 64—66, (1873).
11. Note on some of the iron deposits of Chánda, Central Provinces: *Rec. Geol. Surv. Ind.*, VI, 77—81, (1873) ; *Gazette of India Supplement*, 22 Aug., 1874, pp. 1489—1491.
12. Notes on some of the iron ores of Kumaon : *Rec. Geol. Surv. Ind.*, VII, 15—20, (1874); *Gazette of India Supplement*, 22nd August, 1874, pp. 1466—1468.
13. Note on the raw materials for iron-smelting in the Raniganj field : *Rec. Geol. Surv. Ind.*, VII, 20—30, (1874) ; *Gazette of India Supplement* 22nd Aug., 1874, pp. 1474—1481.
14. Petroleum in Assam : *Rec. Geol. Surv. Ind.*, VII, 55—58, (1874).
15. Second note on the materials for iron manufacture in the Raniganj coal-field : *Rec. Geol. Surv. Ind.*, VII, 122—124, (1874).
16. Manganese ore in the Wardha coal-field : *Rec. Geol. Surv. Ind.*, VII, 125—126, (1874).
17. Note upon the subsidiary materials for artificial fuel : *Rec. Geol. Surv. Ind.*, VII, 160—163, (1874).
18. Trials of Raniganj fire-bricks : *Rec. Geol. Surv. Ind.*, VIII, 18—20, (1875).
19. On the relation of the fossiliferous strata of Maléri and Kota, near Sironcha, Central Provinces: *Rec. Geol. Surv. Ind.*, IX, 86, (1876).
20. The Wardha Valley coal-field : *Mem. Geol. Surv. Ind.*, XIII, 1—154, (1877).
21. Borings for coal in India : *Rec. Geol. Surv. Ind.*, X, 92—97, (1877).
22. Geology of the Upper Godaveri Basin between the Rivers Wardha and the Godavery, near Sironcha : *Rec. Geol. Surv. Ind.*, XI, 17—30, (1878).
23. Note on a trip over the Milam Pass, Kumaon : *Rec. Geol. Surv. Ind.*, XI, 182—187, (1878).

Hughes, T. W. H.,—cont.

24. Satistics of coal importation into India: *Rec. Geol. Surv. Ind.*, XII, 83—87, (1879).

25. Notes on the S. Rewah Gondwana Basin : *Rec. Geol. Surv. Ind.*, XIV, 126—138 and 311—320, (1836).

26. Notes on Mining Records, &c.: *Rec. Geol. Surv. Ind.*, XIV, 185—190, (1881).

27. Note on the Umaria coal-field : *Rec. Geol. Surv. Ind.*, XV, 169—175, (1882).

28. Further notes on the Umaria coal-field: *Rec. Geol. Surv. Ind.*, XVI, 118—121, (1883).

29. Additional notes on the Umaria coal-field: *Rec. Geol. Surv. Ind.*, XVII, 146—150, (1884).

30. The Southern Coal-fields of the Rewah Góndwána Basin, Umaria, Korár, Johilla, Sohágpur, Kúrásia, Koréágarh, Jhilmili : *Mem. Geol. Surv. Ind.*, XXI, 137—249, (1885).

Humboldt, F. H. Alexander von.

1. Sur les chaînes et les volcans de l'intérieur de l'Asie, et sur une nouvelle éruption dans les Andes: *Journ. de. Géol.*, II, 136—173, (1830); *Edin. New Phil. Jour.*, X, 227—240, XI, 145—159, (1831); *Annal. der Phys. u. Chem.*, XVIII, 1—18, (1830), XXIII, 294—301, (1831).

2. Recherches sur les systèmes de montagnes et les volcans de l'intérieur de l'Asie : *Ann. de Chémie*, XLV, 208—215, 337—348, (1830) ; *Glean. Sci.*, III, 330—332, (1831).

3. Fragmens de géologie et de climatologie asiatiques. 2 vols., 8°, Paris, 1831 ; German translation, 8°, Berlin, 1832.

4. Asie Centrale. Recherches sur les chaînes de montagnes et la climatologie comparée. 3 vols., 8°, Paris, 1843 ; German translation, 2 vols., 8°, Berlin, 1844.

Hume, A. O. *and* Henderson, G.   *See* HENDERSON, G. *and* HUME, A. O.

Hunter, A.

1. On the Improvements which have been made in the Pottery of India, &c. : *Ind. Jour. Arts Sci.*, I, 93—105, (1850).

2. On the Pottery of India, showing the various substances required in the different branches of the Art: *Ind. Jour. Arts Sci.*, I, 159—168, (1850).

3. On the indications of Coal in India : *Ind. Jour. Arts Sci.*, I, 208—213, (1850).

4. List of articles from the Animal, Vegetable and Mineral Kingdoms, the Produce of the Madras Presidency : *Ind. Jour. Arts Sci.*, I, 241—261, (1850).

5. Report, &c., upon the Mineral Products of the Madras Presidency sent to the Great National Exhibition of 1851 : *Ind. Jour. Arts Sci.*, I, 431—444, (1851).

6. Lectures on the uses and applications of Plaster of Paris or Gypsum to the Arts and Manufactures : *Ind. Jour. Arts Sci.*, I, 514—524, 580—584, (1851).

Hunter, A.,—cont.

7. Geological excursions near Madras: *Ind. Jour. Arts Sci.*, I, 568—580, (1851).

8. The Resources of the Madras Presidency likely to be done justice to ere long: *Ind. Jour. Arts Sci.*, I, 636—640, (1852).

9. Polishing slate: *Ind. Jour Arts Sci.*, I, 674, (1852).

10. The Fossil Records of creation in Southern India: *Ind. Jour. Arts Sci.*, 2nd series, I, 1—9, 41—45, (1856); 137—140, (1857).

11. Geological excursion at Bangalore: *Ind. Jour. Arts Sci.*, 2nd series, I, 25—27, (1856).

12. Report on a collection of clays, &c., from Debrughur, Upper Assam: *Jour. As. Soc. Beng.*, XXIV, 726—728, (1855).

13. Abstract of a ten years' correspondence on the Resources of India: *Ind. Jour. Arts Sci.*, 2nd series, I, 50—75, (1856).

14. On Coal in South India: *Ind. Economist*, II, 184—187, 210, (1871).

15. Report on the search made in several quarters in Southern India for indications of Coal: *Ind. Economist*, III, 75—76, (1871).

Hunter, W.

1. Analysis of the water drawn from a well in Fort William at the depth of 70 ft. from the surface: *Glean. Sci.*, 103—105, (1829).

Hutchinson, C. H.

1. Report on the new Tenasserim coal-field: *Jour. As. Soc. Beng.*, VIII, 390—391, (1839).

Hutton, T.

1. Journal of a trip to the Burenda Pass in 1836: *Jour. As. Soc. Beng.*, VI, 901—938, (1837).

2. Journal of a trip through Kunawar, Hungrung, and Spiti, undertaken in the year 1838, under the patronage of the Asiatic Society of Bengal, for the purpose of determining the geological formation of the district: *Jour. As. Soc. Beng.*, VIII, 901—949, (1839); IX, 489—513, 555—581, (1840).

3. Geological Report on the Valley of the Spiti and of the route from Kotghur: *Jour. As. Soc. Beng.*, X, 198—229, (1841).

4. Remarks on the Calcutta Delta: *Cal. Jour. Nat. Hist.*, II, 542—560, (1842).

5. Notes on the Geology and Mineralogy of Afghanistan: *Cal. Jour. Nat. Hist.*, VI, 562—611, (1846).

Huxley, J.

1. Sketch of the geology of the Bhartpur District :—*Glean. Sci.*, II, 143—147, (1830).

Huxley, T. H.

1. On some Reptilian Remains from Bengal: *Quart. Jour. Geol. Soc.* XVII, 362, (1861).

Huxley, T. H.,—cont.

2. The Vertebrate Fossils from the Panchet Rocks : *Pal. Indica,* ser. iv, I, pt. i, (1865).

3. On Hyperodapedon : *Quart. Jour. Geol. Soc.,* XXV, 138—152, (1869).

4. On the Classification of the Dinosauria, with observations on the Dinosauria of the Trias ; Dinosauria from the Trias of the Ural Mountains and India : *Quart. Jour. Geol. Soc.,* XXVI, 48, (1870).

I

Impey, E.

1. Memoir on the physical character of the Nerbudda River and Valley, &c., also a descriptive detail of the mineral resources of the Nerbudda Valley, and an analysis of the past correspondence of Government on the subject of the coal-beds in its vicinity : *Sel. Rec. Bo. Gov.,* XIV, 1—37, (1854).

2. Description of the caves of Bagh in Rath : *Jour. Bo. As. Soc.,* V, 543, (1856).

3. Discovery of Ammonitiferous Limestone near Jeysulmere in the Great Desert : *Jour. Bo. As. Soc.,* VI, 161—62, (1862).

Ince, R.

1. Hot Springs of Chittagong : *Jour. As. Soc. Beng.,* XIV, *Proc.,* p. xxiii, (1845).

Indo-China.

1. Miscellaneous papers relating to Indo-China. 8°. London. 2 vols., 1886. Contains FORLONG and FRASER, No. 1 ; J. R. LOGAN, No. 1 ; J. LOW, No. 2 ; G. B. TREMENHEERE, Nos. 2, 4, 6, 7 ; A. URE, No. 2 ; T. WARD, No. 3.

2. Do., 2nd series ; 2 vols. 8°. London, 1887 ; contains J. R. LOGAN, No. 6.

Inverarty, J. D.

1. Report on the Rise, Progress, and Results of the late flood or overflow of the Indus which endangered the towns of Shikarpoor and Jacobabad : *Trans. Bo. Geog. Soc.,* XVI, 48—55, (1863).

Iron.

1. Iron Works at Firozpur, [by A. E.] : *Glean. Sci.,* III, 327-328, (1831).

2. Presentation of specimens of Iron Ore from Sambhalpur : *Jour. As. Soc. Beng.,* I, 250, (1832).

3. Presentation of specimens of Ore and Iron and Steel in various stages of manufacture according to the native processes ; from Salem, Madras ; *Jour. As. Soc. Beng.,* I, 250, (1832).

4. Papers regarding the Forests and Iron mines in Kumaon : *Sel. Rec. Gov. Ind.,* VIII, (Supplement), 1856.

5. Reports on the Survey of the Mineral Deposits in Kumaon and on the smelting operations experimentally conducted at Dechouree : *Sel. Rec. Gov. Ind.,* XVII, (1856).

Irvine, Dr. R. H.
1. Some account of the general and medical topography of Ajmeer. 8°. Calcutta, 1841.

2. A few observations on the probable results of a scientific research after Metalliferous deposits in the Sub-Himalayan range around Darjeeling. *Jour. As. Soc. Beng.*, XVII, pt. i, 137, (1848).

Irwin.
1. Memoir on the Soil, Produce, and Husbandry of Afghanistan and the neighbouring countries: *Jour. As. Soc. Beng.*, VIII, 745—804, 869—900, 1005—1015, (1839); IX, 33—65, 189—197, (1840).

J

J. D. G.
1. The Temperature of water in wells: *Glean Sci.*, II, 131—132, (1830).

Jack, A.
1. Extracts from letters relating to the country between Neemuch and Ferozepore: *Calcutta Jour. Nat. Hist.*, I, 555—557, (1841).

Jack, W.
1. Notice respecting the rocks of the Island of Penang and Singapur, Straits of Malacca; part of a letter to H. T. Colebrooke: *Geol. Trans.*, 2nd series, I, 165, (1824).

Jackson, W.
1. On the Iron Works of Beerbhoom: *Jour. As. Soc. Beng.*, XIV, 754-755, (1845).

2. Notice of two heads found in the Northern Districts of the Punjab: *Jour. As. Soc. Beng.*, XXI, 511, (1852).

Jacob, A. A.
1. Reconnoissance of the Nerbudda Valley in Central India: *Jour. Geol. Soc. Dublin*, VI, 183, (1854).

2. Report on the Iron and Coal Districts of the Nerbudda Valley from Poonassa to Jubbulpore: *Sel. Rec. Bo. Gov.*, new series, IX, 42—48, (1855); XIV, 136—141, (1855).

Jacob, Le Grand.
1. Report on the Iron of Kattywar, its comparative value with British metal, the mines and methods of smelting the ore: *Jour. Roy. As. Soc.*, VII, 98—104, (1843); *Sel. Rec. Bo. Gov.*, XXXVII, 465—471, (1856).

2. Letter prefacing extract from journal of a trip from Sind to Kutch in 1852: *Trans. Bo. Geog. Soc.*, XVI, pp. xxxi-xxxiv, (1863).

3. Extract from Journal of a trip to Sind from Kutch in 1852: *Trans. Bo. Geog. Soc.*, XVI, 22—29, (1863).

4. Extracts from a journal kept during a tour made in 1851 through Kutch, giving some account of the Alum Mines of Murrh, and of changes effected in 1844 by a series of Earthquakes that appear hitherto to have escaped notice: *Trans. Bo. Geog. Soc.*, XVI, 56—66, (1863).

Jacquemont, Victor.

1. Voyage dans l'Inde, pendant les années 1828 à 1832, 4 vols. Text, 2 vols. Atlas.   4°, Paris, 1841.

Jameson, R.

1. On the Graphite, or Black Lead, of Ceylon : *Edin. New Phil. Jour.*, XIII, 346, (1832).

Jameson, W.

1. Remarks on the geology, &c., of the country extending between Bhar and Simla : *Jour. As. Soc. Beng.*, VIII, 1037—1056, (1839).

2. Extract from a letter to Mr. Clerk [Minerals from Kalabagh] : *Jour. As. Soc. Beng.*, XI, 1—3, (1842).

3. First Report by Dr. Jameson of his deputation by Government to examine the effects of the great Inundation of the Indus : *Jour. As.' Soc. Beng.*, XII, 183—225, (1843).

Jamieson, H. W.

1. On Beekite from the Punjab : *Geol. Mag.*, 2nd Decade, VI, 284—286, (1879).

Jamieson, T. F.

1. On the Parallel Roads of Glen Roy, and their place in the History of the Glacial period : *Quart. Jour. Geol. Soc.*, XIX, 235—259, (1863). [Reference to glacial deposits in Central Asia, pp. 257-258.]

Jenkins, F.

1. An account of some minerals collected at Nagpur and its vicinity with remarks on the geology of that part of the country : *As. Res.*, XVIII, pt. 2, 195—215, (1833); *Glean. Sci.*, I, 226-427, (1829).

2. Further discovery of coal beds in Assam *Jour. As. Soc. Beng.*, IV, 704—706, (1835).

3. Announcement of two new sites of coal in Assam : *Jour. As. Soc. Beng.*, VII, 169—170, (1838).

4. [Letters on the discovery of Coal of a very superior description, in a new situation in Upper Assam] : *Cal. Jour. Nat. Hist.*, VII, 213—215, 368—369, (1847).

Jenkins, H. M.

1. On some Tertiary Mollusca from Mount Sela, in the Island of Java (with a note on the nummulitic formation of India) : *Quart. Jour. Geol. Soc.*, XX, 45—66, (1864).

Jenkins, R.

1. Report on the Territories of the Rajah of Nagpore ; submitted to the Supreme Government of India.   4°. Calcutta, 1827.

2. Announcement of two new sites of coal in Assam : *Jour. As. Soc. Beng.*, VII, 169, (1838.)

3. Note respecting the Jaipur [Assam] coal : *Jour. As. Soc. Beng.*, VII, 368, (1838).

Jervis, H.
1. Narrative of a journey to the falls of the Cavery: with an historical and descriptive account of the Neilgherry Hills. 8°. London, 1834.

Jervis, T. B.
1. On slate quarries in the Western Ghauts: *Jour. As. Soc. Beng.*, I, 514—515, (1832).

Jessop & Co.
1. Note on the smelting of the Iron Ore of the District of Burdwan: *Jour. As. Soc. Beng.*, VIII, 683—684, (1839).

Johnston, T. M. H.
1. Coal in the Nizam's territory [Berars]: *Ind. Economist*, III, 177, (1872).

Jones, E. J.
1. Notes on the Kashmir earthquake of 30th May, 1885: *Rec. Geol. Surv. Ind.*, XVIII, 153—155, (1885).
2. Report on the Kashmir earthquake of 30th May, 1885: *Rec. Geol. Surv. Ind*, XVIII, 221—227, (1885).
3. The Southern coal-fields of the Sátpura Gondwána basin (1887): *Mem. Geol. Surv. Ind.*, XXIV, pt. i, 1—58, (1887).
4. On some Nodular stones obtained by trawling off Colombo in 675 fathoms of water: *Jour. As. Soc. Beng.*, LVI, pt. ii, 209—212, (1887).
5. Notes on Upper Burma: *Rec. Geol. Surv. Ind.*, XX, 170—194, (1887).

Jones, S.
1. Some particulars regarding the Mineral Productions of Bengal: *Glean. Sci.*, I, 281—286, (1829).
2. Description of the North-West Coal District, stretching along the River Damuda from the neighbourhood of Jeria, or Juriagerh, to below Sanampur in the Pergunnah of Sheargerh, forming a line of about sixty-five miles: *As. Res.*, XVIII, pt. i, 163—170, (1833); *Glean. Sci.*, I, 261—263, (1829).
3. Remarks on Cancar: *Glean. Sci.*, I, 365—367, (1827).

Jones, T. Rupert.
1. Notes on Fossil *Cypridæ* from Nagpur: *Quart. Jour. Geol. Soc.*, XVI, 186—189, (1860).
2. A monograph of the Fossil *Estheriæ*: *Pal. Soc.*, 4°. London, 1862.
3. On Fossil *Estheriæ* and their distribution: *Quart. Jour. Geol. Soc.*, XIX, 140—157, (1863).

Jumna River.
1. Reports on the Jumna River: *Sel. Rec. Gov., N.-W. P.*, new series, IV, 437—495, (1868).

K

Kalikishen, Raja, *and* Prinsep, J.
1. Oriental accounts of the Precious Minerals (translated by Raja Kalikishen; with remarks by James Prinsep): *Jour. As. Soc. Beng.*, I, 353—363, (1832).

Kankar.

1. On the Production of Cancar: *Glean. Sci.*, I, 247, (1829).

Kasia Hills.

1. Some account of the Casiah Hills [signed F.]: *Glean. Sci.*, I, 252—255, (1829).

2. Excursion to the Chirrapunji Hills [signed C.]: *Glean. Sci.*, III, 172—174 (1831).

3. On the manufacture of the Sylhet Lime: *Glean. Sci.*, II, 61—63, (1830).

Kattiawar.

1. Correspondence relative to geological action on the coast of—: *Trans. Bo. Geog. Soc.*, XVIII, pp. lvi—lxiv, lxix—lxxv, and lxxxv—ci, (1868).

Kaye, C. T.

1. Observations on the Fossiliferous beds near Pondicherry and in the District of South Arcot: *Mad. Jour. Lit. Sci.*, XII, 37—43, (1840); *Cal. Jour. Nat. Hist.*, II, 225—230, (1842).

2. Further observations on the fossiliferous beds near Pondicherry, in continuation of a paper which appeared in the Madras Journal of Literature and Science for July, 1840: *Cal. Jour. Nat. Hist.*, II, 231—237 (1842).

3. On fossils discovered in rocks in Southern India: *Proc. Geol. Soc.*, III 792—793, (1842).

4. Observations on certain fossiliferous beds in Southern India: *Geol. Trans.*, 2nd series, VII, 85—88, (1846); *Proc. Geol. Soc.*, IV, 204—206, (1846).

Keatinge, R. H.

1. Neocomien fossils from Bagh and its neighbourhood, presented by Lieutenant R. H. Keatinge: *Jour. Bo. As. Soc.*, V, 621—623, (1857).

2. [Account of the Cretaceous beds of Bagh]: *Jour. As. Soc. Beng.*, XXVII, 117—123, (1858).

3. Record of the occurrence of Earthquakes in Assam during 1877: *Jour. As. Soc. Beng.*, XLVII, pt. ii, 4—9, (1878).

Keatinge, R. H., *and* Evans.

1. Report on a passage made on the Nurbudda River from the Falls of Dharee to Mundlaisir, by Lieutenant Keatinge, and of a similar passage from Mundlaisir to Baroach by Lieutenant Evans: *Jour. As. Soc. Beng*, XVI, 1104—1112, (1847).

Keene, H. G.

1. Note on the stone industries of Agra: 8°. Mirzapore, 1873.

Kelaart, E. F.

1. Notes on the geology of Ceylon. Laterite formation,—Fluviatile deposits of Newera Elia: *Edin. New Phil. Jour.*, LIV, 28—36, (1852).

Kelsall, J.

1. Geology and Mineralogy [of the Bellary District]: *Bellary Manual.* 8°. Madras, 1872, pp. 90—97.

Kennedy, A.
1. Notice regarding the working and polishing of granite in India : *Edin. Phil. Jour.*, IV, 349—351, (1820).

2. Notice respecting the working and polishing of granite in India : *Edin. Jour. Sci.*, IV, 281—282, (1826).

Kennedy, J. P.
1. Report on the Iron and Coal Districts of the Narbadda Valley, from Poonassa to Jubbulpore : *Sel. Rec. Bo. Gov.*, new series, IX, 42—48, (1855).

Kennedy, R. H.
1. Extract of a letter on the Cornelians, &c., of Guzeratte : *Trans. Med. Phys. Soc. Calcutta*, III, 425—428, (1827).

Kerr.
1. On copper ore from Nellore : *Jour. As. Soc. Beng.*, II, 94—95, (1833).

King, J. S.
1. Descriptive and Historical account of the British Outpost of Perim, Straits of Babelmandeb : *Sel. Rec. Bo. Gov.*, new series, XLIX, (1877).

King, W.
1. On the occurrence of Crystalline limestone in the vicinity of Trichinopoly : *Mad. Jour. Lit. Sci.*, XX, (new series, IV), 272—274, (1858).

2. An Account of the Kolymullays, one of the mountain masses in the Salem District of the Madras Presidency : *Mad. Quart. Jour.*, VIII, 266-282, (1865) ; *Mad. Jour. Lit. Sci.*, 3rd series, I, pt. ii, 63—106, (1866).

3. Notes on the occurrence of stone implements in the North Arcot District : *Mad. Jour. Lit. Sci.*, 3rd series, pt. ii, 36—42, (1866).

4. An account of parts of the Nullamullays, a range of mountains in the Kurnool District : *Mad. Jour. Lit. Sci.*, 3rd series, pt. ii, 63—107, (1866).

5. [On stone implements from Madras] : *Proc. As. Soc. Beng.*, 1867, pp. 139—142.

6. On the Kuddapah and Kurnool formations : *Rec. Geol. Surv. Ind.*, II, 5—10, (1869).

7. Notes on a traverse of parts of the Kummummet and Hannamconda Districts in the Nizam's Dominions : *Rec. Geol. Surv. Ind.*, V, 46—55, (1872).

8. On the Kadapah and Karnul formations in the Madras Presidency : *Mem. Geol. Surv. Ind.*, VIII, pt. 1, 1—291, (1872).

9. Notes on a new coal-field in the south-eastern part of the Hyderabad (Deccan) territory : *Rec. Geol. Surv. Ind.*, V, 65—69, (1872).

10. Note on a possible field of coal-measures in the Godávari District, Madras Presidency : *Rec. Geol. Surv. Ind.*, V, 112—114, (1872).

11. Correction regarding the supposed Eozoönal limestone of Yellambile : *Rec. Geol. Surv. Ind.*, V, 122, (1872).

12. Note on the Barákars (coal-measures) in the Bedadanole field, Godavari District, Madras Presidency : *Rec. Geol. Surv. Ind.*, VI, 57—59, (1873).

King, W.—cont.

13. Note on the progress of geological investigation in the Godávari District, Madras Presidency: *Rec. Geol. Surv. Ind.*, VII, 158—160, (1874).

14. Preliminary note on the gold-fields of south-east Wynaad, Madras Presidency: *Rec. Geol. Surv. Ind.*, VIII, 29—45, (1875).

15. Note on the rocks of the Lower Godávari: *Rec. Geol. Surv. Ind.*, X, 55—63, (1877).

16. [Geology of Cuddapah District]: *Cuddapah Manual.* 8°. Madras, 1875, pp. 16—29.

17. Note on the progress of the Gold Industry in the Wynaad, Nilghiri District, Madras Presidency: *Rec. Geol. Surv. Ind.*, XI, 236—246, (1878).

18. Additional notes on the geology of the Upper Godavari Basin, near Sironcha: *Rec. Geol. Surv. Ind.*, XIII, 13—25, (1880).

19. On the Artesian wells at Pondichery: *Rec. Geol. Surv. Ind.*, XIII, 113—152, (1880).

20. Additional notes on the Artesian wells at Pondichery: *Rec. Geol. Surv. Ind.*, XIII, 194—197, (1880).

21. Des puits artésiens à Pondichéry, et la possibilité de découvrir de sources semblables à Madras. 8°. Pondichéry, (1880).

22. The gneiss and transition rocks and other formations of the Nellore portion of the Carnatic: *Mem. Geol. Surv. Ind.*, XVI, pt. ii, 109—194, (1880).

23. The Upper Gondwanas and other formations of the coastal region of the Godavari District: *Mem. Geol. Surv. Ind.*, XVI, pt. iii, pp. 195—264, (1880).

24. Geology of the Pranhita Godavari Valley: *Mem. Geol. Surv. Ind.*, XVIII, 151—311, (1881).

25. General sketch of the geology of Travancore: *Rec. Geol. Surv. Ind.*, XV, 87—93, (1882).

26. The Warkili beds and reported associated deposits at Quilon: *Rec. Geol. Surv. Ind.*, XV, 93—102, (1882).

27. Record of borings for coal at Beddadanol: *Rec. Geol. Surv. Ind.*, XV, 202—207, (1882).

28. The Singareni coal-field and others adjacent to, or in, the Madras Presidency: 8°. Madras, (1883).

29. Considerations on the smooth water anchorages or mud banks of Narrakal and Alleppy on the Travancore coast: *Rec. Geol. Surv. Ind.*, XVII, 14—27, (1884).

30. On the selection of sites for borings in the Raigarh-Hingir coal-field; first notice: *Rec. Geol. Surv. Ind.*, XVII, 123—130, (1884).

31. Notes on auriferous sands of the Subansiri River; Pondicherry lignite; and phosphatic rocks at Masuri: *Rec. Geol. Surv. Ind.*, XVII, 192—199, (1884).

32. Sketch of the progress of geological work in the Chattisgarh Division of the Central Provinces: *Rec. Geol. Surv. Ind.*, XVIII, 169—200, (1885).

33. Geological sketch of the Vizagapatam District, Madras: *Rec. Geol. Surv. Ind.*, XIX, 143—157, (1886).

King W.,—cont.

34. Memorandum on the Malanj Khandi copper ore : *Rec. Geol. Surv. Ind.*, XIX, 165-166, (1886).

35. Boring exploration in the Chhattisgarh coal-fields : *Rec. Geol. Surv. Ind.*, XIX, 210—234, (1886) ; XX, 194—203, (1887).

King, W., *and* **Foote, R. B.**

1. On the geological structure of parts of the districts of Salem, Trichinopoly, Tanjore and South Arcot, in the Madras Presidency, (being the area included in sheet 79 of the Indian Atlas) : *Mem. Geol. Surv. Ind.*, IV, 223—386, (1864).

Kirkpatrick.

1. An account of the kingdom of Nepaul : 4°. London, 1811.

Kittoe, M.

1. Section of a Hill in Cuttack supposed to be likely to contain Coal : *Jour. As. Soc. Beng.*, VII, 152—153, (1838).

2. Journal of a tour in the Province of Orissa [coal] : *Jour. As. Soc. Beng.*, VII, 679—685, 1060—1063, (1838).

3. Report on the Coal and Iron mines of Talcheer and Ungool, &c., &c.; *Jour. As. Soc. Beng.*, VIII, 137—144, (1839).

4. Account of a Journey from Calcutta, *viâ* Cuttack and Pooree to Sumbulpur, and from thence to Mednipur through the Forests of Orissa :— *Jour. As. Soc. Beng.*, VIII, 367—383, 474 —480, 606—620, 671—681, (1839).

Klaproth, M. H.

1. Sur le spath adamantin. [Transl.] : *Ann. de Chémie*, I, 183—187, (1789).

2. Analyse du Sapphir Oriental. [Transl.] : *Jour. des Mines*, III, No. xv, 51—56, (1796).

3. Analyse de l'oeil de chat. [Transl.] : *Jour. des Mines*, IV, No. XXIII, 9—15, (1796).

4. Analytical Essays towards promoting the chemical knowledge of mineral substances, translated from the German. 8°. London, 1801. Contains Experiments on the Adamantine Spar [Corundum] of Bengal, 64—70 ; Examination of the Oriental [Ceylon] Sapphire, 71—77 ; Examination of the cat's eye from Ceylon and the coast of Malabar, 78—84 ; Examination of the Jargon [Zircon] of Ceylon, 175—194 ; Examination of Spinell [from Ceylon], 316—324 ; Examination of the Oriental Garnet, 334—337.

5. Chemische Untersuchung des Zirkons aus den nördlichen Circars : *Jour. Chemie. u. Physik*, IV, 386—389, (1807).

Knight.

1. Diary of a Pedestrian in Cashmere and Tibet. 8°. London, 1863.

Knight, R. C. *and* **Postans, J.**

1. Reports on the Manchur Lake, and Aral and Narra Rivers : *Jour. Roy. As. Soc.*, VIII, 381—389, (1846).

Kumaon.

1. Official Reports on the Province of Kumaon. 8vo. Calcutta, 1878. Contains J. H. BATTEN, No. 2 ; G. W. TRAILL, NOS. 2, 3.

Kurz, Sulpice.

1. Report on the Vegetation of the Andaman Islands [contains some petrographical information). Flscp. Calcutta, 1870.

L

Lalor, J.

1. A report on Dhur Yaroo, in the Shikárpur Collectorate : *Trans. Bo. Geog. Soc.,* XVII, 302—320, (1865).

2. On the hill districts to the south-west of Mehur in Sind : *Trans. Med. Phys. Soc. Bombay,* VI, 271—283, (1860).

3. On the composition of the liquid disengagea trom Mud volcanoes, &c., in Sind : *Trans. Med. Phys. Soc. Bombay,* VIII, pp. xi-xii, (1862).

4. Rough notes, showing outline of the country between Kurrachee and Gwadel : *Trans. Bo. Geog. Soc.,* XVI, 99—115, (1863).

Langstaff, Dr.

1. Presentation of specimens of sandstone with vegetable impressions from Sikrigali [Rajmehal Hills?]. *Jour. As. Soc. Beng.,* II, 45, (1833).

Lamb.

1. Note on an earthquake at Kamrup on the 11th December, 1872 : *Proc. As. Soc. Beng.,* 1873, p. 65.

La Touche, T. D.

1. The Daranggiri coal-field, Garo Hills : *Rec. Geol. Surv. Ind.,* XV, 175—178, (1882).

2. Note on the Cretaceous coal-measures at Borsora in the Khasia Hills, near Laour in Sylhet : *Rec. Geol. Surv. Ind.,* XVI, 164—166, (1883).

3. Notes on a traverse through the Eastern Khasia, Jaintia, and North Cachar Hills : *Rec. Geol. Surv. Ind.,* XVI, 198—203. (1883).

4. Report on the Langrin coal-field, south-west Khasia Hills : *Rec. Geol. Surv. Ind.,* XVII, 143—146, (1884).

5. Note on the Coal and Limestone in the Dorgrung River near Golaghat, Assam : *Rec. Geol. Surv. Ind.,* XVIII, 31—32, (1885).

6. Notes on the geology of the Aka Hills : *Rec. Geol. Surv. Ind.,* XVIII, 121—124, (1885).

7. Geology of the Upper Dehing Basin in the Singpho Hills : *Rec. Geol. Surv. Ind.,* XIX, 111—115, (1886).

8. Notes on the geology of the Garo Hills : *Rec. Geol. Surv. Ind.,* XX, 40—43, (1887).

Lawder, A. W.

1. Mineralogical statistics of the Kumaon Division : *Rec. Geol. Surv. Ind.,* II, 86—94, (1869).

2. Mineral statistics of Kumaon : *Rec. Geol. Surv. Ind.,* IV, 19—27, (1871).

Leighton, E. W.

1. The Indian Gold Mining Industry; its present condition and its future prospects. 8°. Madras, 1883.

Leith, A. H.

1. Discovery of more organic remains and minerals in the Trap of Bombay: *Jour. Bo. As. Soc.*, VI, 180, (1861).

2. The Town and Island of Bombay; BOMBAY, pp. 15—22, (1878)

Le Mesurier, H. P.

1. [Description of finds of stone celts in Bundelcund]: *Jour. As. Soc. Beng.*, XXX, 81—85, (1861).

Leonard, H.

1. Memorandum on the River Hooghly: *Sel. Rec. Gov. Ind.*, XLV, 1—21, (1864).

2. [Earthquakes in June and July, 1868]: *Proc. As. Soc. Beng.*, 1868, pp. 256—257.

3. [On the earthquake of 1869 in Cachar]: *Proc. As. Soc. Beng.*, 1869, pp. 102—103.

Liebig, G. von.

1. Barren island (Account of a visit to Barren Island in March 1858): *Zeits. Deutch. Geol. Gesel.*, X, 299—304, (1858); *Jour. As. Soc. Beng.* XXIX, 1—10, (186); *Sel. Rec. Gov. Ind.*, XXV, 124—131, (1859).

Lime.

1. On the manufacture of the Sylhet Lime: *Glean. Sci.*, II, 61—63, (1830).

Liston, D.

1. Notes on the distribution of soils in the Goruckpore District: *Cal. Jour. Nat. Hist.*, I, 236—240, (1841).

2. Note regarding the Salts in the soil of the eastern portion of Zilla Goruckpore: *Cal. Jour. Nat. Hist.*, II, 125—126, (1842)

3. Some Memoranda on the geology of Sikkim: *Cal. Jour. Nat. Hist.*, IV, 521—532, (1844).

Lithography.

1. Lithographic printing in Tibet: *Glean. Sci.*, I, 110, (1829).

2. On the rise and progress of the Lithographic Art in India, with a brief notice of the Native Lithographic stones of that Country: *Glean. Sci.*, I, 54—56, (1829).

3. On the application of the Jaisulmer Limestone to the purposes of Lithography: *Glean. Sci.*, III, 107—110, (1831).

Lloyd, H. E. *and* Orlich, L. Von.

1. Travels in India (of Leopold Von Orlich), including Sinde and the Punjab, translated by H. E. Lloyd. 2 vols., 8°. London, 1845.

Lloyd, R.

1. A short notice of the coast line, rivers, and islands adjacent, forming a portion of the Mergui Province, from a late Survey: *Jour. As. Soc. Beng.*, VII, 1027—1038, (1838).

Lloyd, *Sir* W., *and* Gerard, Alex.

1. Narrative of a journey from Caunpore to the Boorendo Pass in the Himalaya Mountains, *vid* Gwalior, Agra, Delhi and Sirhind, by Major Sir William Lloyd ; and an account of an attempt to penetrate by Bekhur to Garoo and the Lake Manasarowara, by Captain Alexander Gerard ; with a letter from the late J. G. Gerard, detailing a visit to the Shaitool and Boorendo Passes, for the purpose of determining the line of perpetual snow on the southern face of the Himalaya, &c. Edited by George Lloyd. 2 vols., 8°. London, 1840.

Loch, G.

1. Sylhet coal : *Jour. As. Soc. Beng.*, VII, 959—963, (1838).

Logan, J. R.

1. On the Local and Relative Geology of Singapore, including notices of Sumatra, the Malay Peninsula &c. : *Jour. As. Soc. Beng.*, XVI, 519—557, 667—684, (1847) ; Indo-China, No. 1, II, 64—112, (1886).

2. Discovery of coal in Ligor and Kedah, on the West coast of the Malay Peninsula : *Jour. Ind. Archip.*, I, 151—168, (1847).

3. Sketch of the Physical Geography and Geology of the Malay Peninsula : *Jour. Ind. Archip.*, II, 83—138, (1848).

4. Notices of the Geology of the East Coast of Johore : *Jour. Ind. Archip.*, II, 625—631, (1848).

5. Notice of the discovery of coal on one of the islands on the coast of the Malay Peninsula : *Quart. Jour. Geol. Soc.*, IV, 1—2, (1849).

6. The rocks of Pulo Ubin : with some remarks on the formation and structure of Hypogene rocks and on the metamorphic Theory : *Verhand, Batav. Genootsch. vun. Kunst. en Wettenshap.* XXII, (1849) ; Indo-China, No. 2, I, 21—71, (1887).

7. Notices of the Geology of the Straits of Singapore : *Quart. Jour. Geol. Soc.*, VII, 310—314, (1851) ; *Jour. Ind. Archip.*, VI, 179—217, (1852).

Login, T.

1. On the delta of the Irrawaddy : *Cal. Engin. Jour.*, I, 375—376, (1858).

2. Memoranda on the most recent Geological changes of the Rivers and Plains of Northern India, founded on accurate surveys and the Artesian well-boring at Umballa, to show the practical application of Mr. Login's theory of the abrading and transporting power of water to effect such changes : *Quart. Jour. Geol. Soc.*, XXVIII, 186—199, (1872).

Lord, P. B.

Some account of a visit to the plain of Koh-i-Daman, the mining district of Ghorband, and the Pass of Hindu Kúsh, with a few general observations respecting the structure and conformation of the country from the Indus to Kábul : *Jour. As. Soc. Beng.*, VII, 521—537, (1838) : *Cal. Jour Nat. Hist.*, I, 564—574, (1861).

Low, J.

1. Notice of the Phoonga Caves in Junk Ceylon : *Edin. Jour. Sci.*, VII, 57—58, (1827).

Low, J.,—cont.

2. Observations on the Geological Appearances and General Features of portions of the Malayan Peninsula, and of the countries lying betwixt it and 18° North Latitude : *As. Res.* XVIII, pt. i, 128—162, (1833) ; *Glean. Sci.*, I, 223—225, (1829) ; Indo-China, No. 1, I, 179—201, (1886).

3. A brief dissertation on the soil and agriculture of the British settlement of Penang or Prince of Wales' Island, in the Straits of Malacca ; including the Province of Wellesley, on the Malayan Peninsula, with brief references to the Settlements of Singapore and Malacca. 8°. Singapore, 1836.

4. Notes on the geological features of Singapore and some of the Islands adjacent : *Jour. Ind. Archip.*, I, 83—100, (1847).

5. Notes on the coal-deposits which have been discovered along the Siamese coast from Penang to the vicinity of Junk Ceylon : *Jour. Ind. Archip.*, I, 145—149, (1847).

6. Memoranda respecting the Sumatran coal : *Jour. Ind. Archip.*, II, 755—757, (1848).

Lumsden, J. G.

1. [Extract of letter] on the island of Perim [Gulf of Cambay]: *Jour. Bo. As. Soc.*, I, 25—30, (1841).

Lush, C.

1. Geological notes on the Northern Conkan, and a small part of Guzerat and Kattiawar : *Jour. As. Soc. Beng.*, V, 761—767, (1836).

Lushington, G. S.

1. Report on the Government experimental working of the copper mines of Pokree in Ghurwal, with notices of other copper mines : *Jour. As. Soc. Beng.*, XII, 453—472, 769, (1843).

Lydekker, R.

1. List of the fossils collected by A. B. Wynne, in the Salt Range : *Rec. Geol. Surv. Ind.*, VIII, 48—49, (1875).

2. Exhibition of a portion of the lower jaw of *Tetraconodon magnum*, Falconer, from the Sivaliks : *Proc. As. Soc. Beng.*, 1876, p. 172.

3. Description of a cranium of *Stegodon ganesa*, with notes on the sub-genus and allied forms : *Rec. Geol. Surv. Ind.*, IX, 42—49, (1876).

4. Notes on the fossil Mammalian faunæ of India and Burmah : *Rec. Geol. Surv. Ind.*, IX, 86—106, 154, (1876).

5. Notes on the osteology of *Merycopotamus dissimilis: Rec. Geol. Surv. Ind.*, 144—153, (1876).

6. Occurrence of *Plesiosaurus* in India : *Rec. Geol. Sur. Ind.*, IX, 154, (1876).

7. Notes on the geology of the Pir-Panjál and neighbouring districts : *Rec. Geol. Sur. Ind.*, IX, 155—162, (1876).

8. Indian Tertiary and Post-Tertiary vertebrata : *Pal. Indica*, series x.

     Vol. I, Part 2 (1877).—Molar teeth and other remains of Mammalia.

     „   Part 3 (1878).—Crania of Ruminants.

     „   Part 4 (1880).—Supplement to above.

     „   Part 5 (1880).—Siwalik and Narbada Proboscidia.

Lydekker, R.,—cont.

8. Indian Tertiary and Post-Tertiary vertebrata : *Pal. Indica*, series x.—*cont.*

Vol. II, Part 1 (1881).—Siwalik Rhinocerotidæ.

    „   Part 2 (1881).—Supplement to Siwalik and Narbada Proboscidia.

    „   Part 3 (1882).—Siwalik and Narbada Equidæ.

    „   Part 4 (1883).—Siwalik Camelopardalidæ.

    „   Part 5 (1883).—Siwalik Selenodont Suina.

    „   Part 6 (1884).—Siwalik and Narbada Carnivora.

Vol. III, Part 1 (1884).—Additional Siwalik Perissodactyla and Proboscidia.

    „   Part 2 (1884).—Siwalik and Narbada Bunodont Suina.

    „   Part 3 (1884).—Rodents and new Ruminants from the Siwaliks, and synopsis of Mammalia.

    „   Part 4 (1884).—Siwalik Birds.

    „   Part 5 (1884).—Mastodon teeth from Perim Island.

    „   Part 6 (1885).—Siwalik and Narbada Chelonia.

    „   Parts 7 & 8 (1886).—Siwalik Crocodilia, Lacertilia, and Ophidia ; and Tertiary Fishes.

Vol. IV, Part 1 (1886).—Siwalik Mammalia, Supplement I.

    „   Part 2 (1886).—The Fauna of the Karnul Caves.

    „   Part 3 (1887).—Eocene chelonia from the Salt Range.

9. Notices of new and other vertebrata from Indian tertiary and secondary rocks : *Rec. Geol. Surv. Ind.*, X, 30—43, (1877).

10. Notices of new or rare mammals from the Siwaliks : *Rec. Geol. Surv. Ind.*, X, 76—83, (1877).

11. Note on the genera *Chæromeryx* and *Rhagatherium* : *Rec. Geol. Surv. Ind.*, X, 225, (1877).

12. Exhibition of the palate of an Anthropoid ape found in the Siwaliks of Punjab : *Proc. As. Soc. Beng.*, 1878, p. 191.

13. Notes on the geology of Kashmir, Kishtwar, and Pangi : *Rec. Geol. Surv. Ind.*, XI, 30—64, (1878).

14. Notices of Siwalik Mammals : *Rec. Geol. Surv. Ind.*, XI, 64—104, (1878).

15. Indian Pretertiary Vertebrata. The Reptilia and Batrachia : *Pal. Indica*, series iv, I, pt. iii, (1879).

16. Geology of Cashmir, (3rd notice) : *Rec. Geol. Surv. Ind.*, XII, 15—32, (1879).

17. Further notices of Siwalik Mammalia : *Rec. Geol. Surv. Ind.*, XII, 33—52, (1879).

18. On some Siwalik Birds : *Rec. Geol. Surv. Ind.*, XII, 52—57, (1879).

19. Popular guide to the Geological collections in the Indian Museum, Calcutta. No. 1 ; Tertiary Vertebrate animals. 8°. Calcutta, 1879.

20. A Sketch of the History of the Fossil Vertebrata of India : *Jour. As. Soc. Beng.*, XLIX, pt. ii, 8—40, (1880).

21. Geology of Ladak and neighbouring districts : *Rec. Geol. Surv. Ind.*, XIII, 26—59, (1880).

Lydekker, R.,—cont.

22. Teeth of fossil fishes from Ramri Island and the Punjab: *Rec. Geol. Surv. Ind.*, XIII, 59—61, (1880).

23. Geology of part of Dardistan, Baltistan, and neighbouring districts: *Rec. Geol. Surv. Ind.*, XIV, 1—56, (1881).

24. Note on some Siwalik Carnivora: *Rec. Geol. Surv. Ind.*, XIV, 57—66, (1881).

25. Note on some Mammalian fossils from Perim Island: *Rec. Geol. Surv. Ind.*, XIV, 155—156, (1881).

26. Note on some Gondwana Vertebrates: *Rec. Geol. Surv. Ind.*, XIV, 174—178, (1881).

27. Observations on the ossiferous beds of Hundes in Thibet: *Rec. Geol. Surv. Ind.*, XIV, 178—184, (1881).

28. Geology of N.-W. Kashmir and Khagan: *Rec. Geol. Surv. Ind.*, XV, 14—24, (1882).

29. On some Gondwana Labyrinthodonts: *Rec. Geol. Surv. Ind.*, XV, 24—28, (1882).

30. Note on some Siwalik and Jumna mammals: *Rec. Geol. Surv. Ind.*, XV, 28—33, (1882).

31. Note on some Siwalik and Narbada fossils: *Rec. Geol. Surv. Ind.*, XV, 102—107, (1882).

32. Synopsis of the fossil Vertebrata of India: *Rec. Geol. Surv. Ind.*, XVI, 61—93, (1883).

33. Note on the Bijori Labyrinthodont: *Rec. Geol. Surv. Ind.*, XVI, 93—94, (1883).

34. On skull of *Hippotherium antelopinum*: *Rec. Geol. Surv. Ind.*, XVI, 94, (1883).

35. Note on the probable occurrence of Siwalik Strata in China and Japan: *Rec. Geol. Surv. Ind.*, XVI, 158—161, (1883).

36. Note on the occurrence of *Mastodon angustidens* in India: *Rec. Geol. Surv. Ind.*, XVI, 161—162, (1883).

37. Geology of the Cashmir and Chamba territories, and the British District of Khagan: *Mem. Geol. Surv. Ind.*, XXII, (1883).

38. Note on the occurrence of the genus *Lyttonia*, Waag., in the Kuling series of Kashmir: *Rec. Geol. Surv. Ind.*, XVII, 37, (1884).

39. Note on the Distribution in Time and Space of the Genera of Siwalik Mammals and Birds: *Geol. Mag.*, 3rd decade, I, 489—492, (1884).

40. Catalogue of Vertebrate Fossils from the Siwaliks of India, in the Science and Art Museum, Dublin: *Trans. Roy. Dub. Soc.*, 2nd ser., III, 69—86, (1884).

41. Catalogue of the Remains of Siwalik vertebrata contained in the geological Department of the Indian Museum, Calcutta, pt. I, Mammalia. 8°. Calcutta, 1885. Pt. II, Aves, Reptilia, and Pisces, 1886.

42. Note on a second species of Siwalik camel (*Camelus antiquus*, nobis and Falc. et Caut. MS.): *Rec. Geol. Surv. Ind.*, XVIII, 78—79, (1885).

Lydekker, R.,—cont.

43. Note on a third species of *Merycopotamus*: *Rec. Geol. Surv. Ind.*, XVIII, 145-146, (1885).

44. A revision of the Antelopes of the Siwaliks: *Geol. Mag.*, 3rd decade, II, 169—171, (1885).

45. Note on some Siwalik Bones erroneously referred to a Struthoid (*Dromæus* (?) *sivalensis*, Lyd.): *Geol. Mag.*, 3rd decade, II, 237, (1885).

46. The Labyrinthodont from the Bijori group: *Pal. Indica*, series iv, I, pt. iv, (1885).

47. Maleri and Denwa Reptilia and Amphibia: *Pal. Indica*, series iv, I, pt. v, (1885).

48. Catalogue of Fossil Mammalia in the British Museum. 5 Parts. 8°. London, 1885-87.

49. Preliminary note on the mammalia of the Karnul caves: *Rec. Geol. Surv. Ind.*, XIX, 120-122, (1886).

50. Catalogue of Pleistocene and Prehistoric vertebrata: contained in the Geological Department of the Indian Museum, Calcutta. 8°. Calcutta, 1886.

51. On a new Emydine Chelonian from the Pliocene of India: *Quart. Jour. Geol. Soc.*, XLII, 540-541, (1886).

52. Description of a jaw of *Hyotherium*, from the pliocene of India: *Quart. Jour. Geol. Soc.*, XLIII, 19-22, (1887).

53. On certain Dinosaurian Vertebræ from the Cretaceous of India and the Isle of Wight: *Quart. Jour. Geol. Soc.*, XLIII, 156-160, (1887).

Lyman, Benjamin Smith.

1. General Report on the Punjab oil lands. Flsc. Lahore, 1870.

2. Topography of the Punjab oil Regions: *Trans. Am. Phil. Soc.*, XV, 1—14, (1872).

M

McClelland, J.

1. Report on the physical condition of the Assam Tea plant, with reference to Geological Structure, Soils and Climate: *Mad. Jour. Lit. Sci.*, VI, 423-444, (1837).

2. Notice of some Fossil Impressions occurring in the Transition Limestone of Kemaon: *Jour. As. Soc. Beng.*, III, 628—630, (1834).

3. Some inquiries in the Province of Kemaon, relative to geology, and other branches of Natural Science. 8° Calcutta, 1835.

4. On the fossil shells of the Kasya hills: *Jour. As. Soc. Beng.*, IV, 520, (1885).

5. Catalogue of the geological specimens from Kemaon, presented to the Asiatic Society: *Jour. As. Soc. Beng.*, VI, 653-663, (1837).

6. On the Difference of Level in Indian coal-fields and the causes to which this may be ascribed: *Jour. As. Soc. Beng.*, VII, 65-82, (1838); also No. 10, pp. 22-46.

7. [Note on a specimen of mud brought up from a depth of 200 fathoms in the Swatch]: *Jour. As. Soc. Beng.*, VII, 369, (1838).

8. On the genus Hexaprotodon of Dr. Falconer and Captain Cautley: *Jour. As. Soc. Beng.*, VII, 1038—1047, (1838).

McClelland, J.,—cont.

9. On the geology of part of Upper Assam : *Proc. Geol. Soc.*, II, 566-588, (1838); *Bibl. Univ.*, XIV, 189-192, (1838).

10. Reports of a committee for investigating the resources of India with reference to coal and iron. 8ᶜ. Calcutta, 1838, pp. 9-96. See COAL, No. 2 ; contains No. 6 *supra*.

11. Report of the coal committee. 8°. Calcutta, 1840; *Mad. Jour. Lit. Sci.*, XI, 355-371, (1840).

12. On Cystoma a new genus of fossil Echinidea [from Cherra Poonji] : *Cal. Jour. Nat. Hist.*, I, 155-187, (1841).

13. Stone and Marble quarries at Mirzapore : *Cal. Jour. Nat. Hist.*, 429-430, (1841).

14. Remarks on the Deposits of the Calcutta Basin : *Cal. Jour. Nat. Hist.*, I, 452-458, (1841).

15. Extracts from a letter to Government on Capt. G. B. Tremenheere's report on the Tin of Mergui : *Jour. As. Soc. Beng.*, XI, 25, (1842).

16. Revised notes on the Fossils discovered by Messrs. Kaye and Cunliffe at Seedapett : *Cal. Jour. Nat. Hist.*, II, 238-244, (1842).

17. Notice of a Fossil Fish, the supposed Rana diluvii testis, or "Fossil Batrachian" of Dr. Cantor : *Cal. Jour. Nat. Hist.*, III, 83-87, (1843).

18. Report of a committee for the investigation of the coal and mineral resources of India for May, 1845. Flscp. Calcutta, 1846.

19. Notices regarding some fossil specimens from the neighbourhood of Lullutpore : *Jour. Agri-Hort. Soc. Ind.*, VI, pt. ii, 5—6, (1848).

20. Report of the Geological Survey for 1848-1849. 4°. Calcutta, 1850.

21. Note on the Discharge of Water by the Irawaddy : *Jour. As. Soc. Beng.*, XXII, 480—484, (1853).

22. Report on the Southern Forests of Pegu : *Sel. Rec. Gov. Ind.*, IX, 6—26, (1855).

23. Analytical Report on specimens of coal from the Nicobar : *Sel. Rec. Gov. Ind.*, LXXVII, 28, (1870).

McClelland, J., *and* Griffith, W.

1. Journals of Travels in Assam, Burma, Bootan, Afghanistan and the neighbouring countries : posthumous papers arranged by Dr. J. M. McClelland. 8.° Calcutta, 1847 ; W. GRIFFITH, No. 4.

Mackenzie, Colin.

1. Account of the Pagoda at Jerwuttum : *As. Res.*, V, 303—314, (1794). [Iron and Diamonds in Karnul.]

McKennie, J.

1 Eight years' observations upon the effects of the groynes (twenty in number), with which is an attempted exposition of the theory of the Madras surf, submitted to the Commandant and Chief Engineer : *Mad. Jour. Lit. Sci.*, XXI, (new series, V), 342—348, (1859).

Mackeson, F.

1 Report on the route from Seersa to Bahawalpore : *Jour. As. Soc. Beng.*, XIII, 297—313, (1844).

McLagan, R.

1. [Coal and petroleum of the Punjab]: *Jour. Soc. Arts*, XXX, 594, (1882).

2. The rivers of the Punjab : *Proc. Roy. Geog. Soc.*, new series, VII, 705—718, (1885).

McLeod, D.

1. Abstract Report of the Proceedings of the Committee appointed to superintend the Boring Operations in Fort William, from their commencement in December, 1835, to their close in April, 1840 : *Jour. As. Soc. Beng.*, IX, 677—687, (1840).

McLeod, J.

1. Memoranda on the pearl-banks and pearl-fishery, the sea-fishery and the salt-beds, of Sind : *Sel. Rec. Bo. Gov.*, XVII, 699—707, (1855).

McLeod, T. E.

1. Abstract journal of an expedition to Kiang Hung on the Chinese Frontier, starting from Moulmein on the 13th December, 1836 : *Jour. As. Soc. Beng.*, VI, 989—1005, (1837).

2. On the Hotsprings of Palouk, Tenasserim : *Jour. As. Soc. Beng.*, VII, 466—467, (1838).

McLeod, W.

1. Memorandum regarding specimens from Seoni, Chupara : *Jour. As. Soc. Beng.*, VI, 1091—1102, (1837).

McMahon, C. A.

1. The Blaini group and the "Central Gneiss" in the Simla Himalayas : *Rec. Geol. Surv. Ind.*, X, 204—223, (1877).

2. Notes on a tour through Hangrang and Spiti : *Rec. Geol. Surv. Ind.*, XII, 57—69, (1879).

3. Note on the section from Dalhousie to Pangi *via* the Sach pass : *Rec. Geol. Surv. Ind.*, XIV, 305—310, (1881).

4. The Geology of Dalhousie : *Rec. Geol. Surv. Ind.*, XV, 34—51,(1882).

5. On the Traps of Darang and Mandi in the N.-W. Himalayas : *Rec. Geol. Surv. Ind.*, XV, 155—164, (1882).

6. Some notes on the Geology of Chamba : *Rec. Geol. Surv. Ind.*, XVI, 35—42, (1883).

7. On the Basalts of Bombay : *Rec. Geol. Surv. Ind.*, XVI, 42—50, (1883).

8. On the microscopic structure of some Dalhousie rocks : *Rec. Geol. Surv. Ind.*, XVI, 129—144, (1883).

9. On the lavas of Aden : *Rec. Geol. Surv. Ind.*, XVI, 145—158, (1883).

10. On the altered basalts of the Dalhousie region : *Rec. Geol. Surv. Ind.*, XVI, 178—186, (1883).

11. On the microscopic structure of some Sub-Himalayan rocks : *Rec. Geol. Surv. Ind.*, XVI, 186—192, (1883).

12. Notes on the Geology of the Chuari and Sihunta pargannahs of Chamba : *Rec. Geol. Surv. Ind.*, XVII, 34—37, (1884).

13. On the microscopic structure of some Himalayan granites and gneissose granites : *Rec. Geol. Surv. Ind.*, XVII, 53—73, (1884).

McMahon, C. A.,—cont.

14. On the microscopic structure of some Arvali rocks: *Rec. Geol. Surv. Ind.*, XVII, 101—118, (1884).

15. On fragments of slates and schists imbedded in the gneissose granite and granite of the N.-W. Himalayas: *Rec. Geol. Surv. Ind.*, XVII, 168—175, (1884).

16. Some further notes on the geology of Chamba: *Rec. Geol. Surv. Ind.*, XVIII, 79—110, (1885).

17. Notes on the section from Simla to Wangtu and on the Petrological character of the Amphibolites and Quartz Diorites of the Sutlej valley: *Rec. Geol. Surv. Ind.*, XIX, 65—88, (1886).

18. On the microscopic character of some Eruptive rocks from the Central Himalayas: *Rec. Geol. Surv. Ind.*, XIX, 115—119, (1886).

19. Notes on the microscopic structure of some specimens of the Malani rocks of the Aravali region: *Rec. Geol. Surv. Ind.*, XIX, 161—165, (1886).

20. Note on some Indian image stones: *Rec. Geol. Surv. Ind.*, XX, 43—45, (1887).

21. Note on the Microscopic structure of some specimens of the Rajmahál and Deccan traps: *Rec. Geol. Surv. Ind.*, XX, 104—111, (1887).

22. Some notes on the Dolerite of the Chor: *Rec. Geol. Surv. Ind.*, XX, 112—117, (1887).

23. Note on the foliation of the Lizard Gabbro. [Mode of intrusion of the Himalayan Granites]: *Geol. Mag.*, 3rd decade, IV, 74—78, (1887).

24. The Gneissose Granite of the Himalayas: *Geol. Mag.*, 3rd decade, IV, 212—220, (1887).

25. Some Remarks on Pressure metamorphism with reference to the Foliation of the Himalayan gneissose granite: *Rec. Geol. Surv. Ind.*, XX, 203—205, (1887).

26. A list and Index of papers on Himalayan Geology and Microscopic petrology by Col. C. A. McMahon, published in the preceding volumes of the Records of the Geological Survey of India: *Rec. Geol. Surv. Ind.*, XX, 206—214, (1887).

McPherson, J.

1. Table of Mineral springs in British India, with a few Remarks: *Ind. Ann. Med. Sci.*, II, 205—221, (1855).

McPherson, S. C.

1. On the Geology of the Peninsula: *As. Res.*, XVIII, 115—221, (1833).

Madras.

1. Standing Information regarding the Official Administration of the Madras Presidency. 8°. Madras, 1877; Mineral Resources, pp. 239—241.

2. Do. 2nd ed., 8°, Madras, 1879; Mineral resources, pp. 225—230.

Madden, E.

1. Diary of an Excursion to the Shatool and Boorun passes over the Himalaya in September, 1845: *Jour. As. Soc. Beng.*, XV, 79—134, (1846).

2. Notes of an Excursion to the Pindree glacier, in September, 1846: *Jour. As. Soc. Beng.*, XVI, 226—266, 596, (1847).

Madden, E.,—cont.

3. The Turaee and Outer Mountains of Kumaon: *Jour. As. Soc. Beng.*, XVII, pt. i, 349—450, 563—626, (1848).

4. Supplementary notes to "The Turaee and Outer Mountains of Kumaon:" *Jour. As. Soc. Beng.*, XVIII, 603—644, (1849).

Magrath, R. N.

1. Some observations upon Scinde and the river Indus as far up as Bukkur: *Trans. Bo. Geog. Soc.*, May 1839, pp. 25—31.

Malay Peninsula.

1. On the Metalliferous deposits of the Malay Peninsula: *Edin. New Phil. Jour.*, XLV, 332—335, (1848), quoting *Jour.· Ind. Archip.*, II, 102.

Malcolm, *Sir* J.

1. A memoir of Central India, including Malwa, and adjoining provinces: 2 vols., 8°. London, 1823; contains F. DANGERFIELD, No. 1, in appendix, vol. II, pp. 313—349.

Malcolmson, J. G.

1. Notice of the fall of an Aërolite [on 2nd Jan., 1831, at Mangapatnam]: *Glean. Sci.*, III, 389, (1831).

2. Note on Saline deposits in Hyderabad: *Jour. As. Soc. Beng.*, II, 77—79, (1833).

3. [Letter forwarding Geological specimens collected by Dr. Voysey]: *Jour. As. Soc. Beng.*, II, 94, and Plate XX, p. 582, (1833).

4. Fragment of bone from a cave in the neighbourhood of Hyderabad: *Jour. As. Soc. Beng.*, II, 204—205, (1833).

5. On the Fossils of the Eastern Portion of the Great Basaltic District of India: *Geol. Trans.*, 2nd series, V, 537—576, (1840): *Proc. Geol. Soc.*, II, 579—584, (1833—1838): *Mad. Jour. Lit. Sci.*, XII, 58—104, (1840), WESTERN INDIA, 1—47, (1857).

6. Fossil Shells in Hyderabad: *Jour. As. Soc. Beng.*, III, 302—303, (1834).

7. Geology of the South of India: *Mad. Jour. Lit. Sci.*, I, 329—342, (1834).

8. Notes explanatory of a collection of geological specimens from the country between Hyderabad and Nagpur: *Jour As. Soc. Beng.*, V, 96—121, (1836); *Mad. Jour. Lit. Sci.*, IV, 194—218, (1836).

9. On the geology of the Basaltic plateau of India: *Mad. Jour. Lit. Sci.*, VIII, 203—210, (1838).

10. Note on Fossil Plants discovered in the Sandstone rocks at Kamptee, near Nagpur: *Jour. Bo. As. Soc.*, I, 249—251, (1843).

11. On the occurrence of Quicksilver in the Lava rocks of Aden: *Jour. Bo. As. Soc.*, I, 341—344, (1843).

12. [On stratified beds in contact with gneiss in Southern India]: *Cal. Jour. Nat. Hist.*, IV, 107—112, (1844).

13. Note on Lacustrine Tertiary Fossils from the Vindhyan Mountains near Mandoo, and on the period of the elevation of that chain: *Trans. Bo. Geog. Soc.*, V, 368—375, (1844).

14. Account of Aden: *Jour. Roy. As. Soc.*, VIII, 279—292, (1846).

Mallet, F. R.

1. On the gypsum of Lower Spiti, with a list of minerals collected in the Himalayas in 1864: *Mem. Geol. Surv. Ind.,* V, 153—172, (1865).

2. Copper in Bundelkand : *Rec. Geol. Surv. Ind.,* I, 16—17, (1868).

3. On the Vindhyan series as exhibited in the North-Western and Central Provinces of India : *Mem. Geol. Surv. Ind.,* VII, pt. i, 1—129, (1869).

4. On the Geological structure of the country near Aden, with reference to the practicability of sinking Artesian Wells : *Mem. Geol. Surv. Ind.,* VII, pt. iii, 257—284, (1871).

5. Mineralogical notes on the gneiss of South Mirzapur and adjoining country : *Rec. Geol. Surv. Ind.,* V, 18—23, (1872); VI, 42—44, (1873).

6. Geological notes on part of Northern Hazáribagh : *Rec. Geol. Surv. Ind.,* VII, 32—44, (1874).

7. Notes from the Eastern Himalayas : *Rec. Geol. Surv. Ind.,* VII, 53, (1874).

8. On the Geology of the Darjiling district and the Western Duars : *Mem. Geol. Surv. Ind.,* XI, pt. i, 1—96, (1874).

9. Note on coals recently found near Moflong, Khásia Hills : *Rec. Geol. Surv. Ind.,* VIII, 86, (1875).

10. On the coal-fields of the Naga Hills, bordering the Lakhimpur and Sibsagar districts, Assam : *Mem. Geol. Surv. Ind.,* XII, pt. ii, 269—363, (1876).

11. On recent coal explorations in the Darjíling District : *Rec. Geol. Surv. Ind.,* X, 143—148, (1877).

12. Limestone in the neighbourhood of Barákar : *Rec. Geol. Surv. Ind.,* X, 148—152, (1877).

13. On some forms of blowing machine used by the Smiths of Upper Assam : *Rec. Geol. Surv. Ind.,* X, 152—154, (1877).

14. The mud volcanoes of Ramri and Cheduba : *Rec. Geol. Surv. Ind.,* XI, 188—207, (1888) ; Burma, pp. 238—259, (1882).

15. Mineral resources of Ramri, Cheduba, and adjacent islands : *Rec. Geol. Surv. Ind.,* XI, 207—223, (1878) ; Burma, pp. 259—277, (1882).

16. Note on a recent mud eruption in Ramri Island : *Rec. Geol. Surv. Ind.,* XII, 70—72, (1879) ; Burma, 277—281, (1882).

17. On Braunite, with Rhodonite from near Nagpur : *Rec. Geol. Surv. Ind.,* XII, 73—74, (1879).

18. On Pyrolusite, with Psilomelane, occurring at Gosalpur, Jabalpur district : *Rec. Geol. Surv. Ind.,* XII, 99—100, (1879).

19. On Mysorin and Atacamite from Nellore : *Rec. Geol. Surv. Ind.,* XII, 166—172, (1879).

20. On corundum from the Khasi Hills : *Rec. Geol. Surv. Ind.,* XII, 172, (1879).

21. Popular guide to the Geological collections in the Indian Museum, Calcutta. No. 2, Minerals. 8°. Calcutta, 1879.

22. Record of gas and mud eruptions on the Arrakan coast on 12th March, 1879, and in June, 1843 : *Rec. Geol. Surv. Ind.,* XIII, 206—209, (1880) ; Burma, pp. 281—284.

Mallet, F. R.,—cont.

23. On the ferruginous beds associated with the Basaltic rocks of North-Eastern Ulster. [Refers to Laterite]: *Rec. Geol. Surv. Ind.,* XIV, 139—148, (1881).

24. On Cobaltite and Danaite from the Khetri mines: *Rec. Geol. Surv. Ind.,* XIV, 190—196, (1881).

25. On the occurrence of zinc ore (Smithsonite and Blende) in the Karnul district : *Rec. Geol. Surv. Ind.,* XIV, 196, (1881).

26. Notice of a mud eruption at Cheduba : *Rec. Geol. Surv. Ind.,* XIV, 196—197, (1881) ; Burma, p. 285, (1882).

27. On Oligoclase granite at Wangtu on the Sutlej : *Rec. Geol. Surv. Ind.,* XIV, 238—240, (1881).

28. On native antimony from Pulo Obin, near Singapore : *Rec. Geol. Surv. Ind.,* XIV, 303—304, (1881).

29. On Turgite from near Juggiapet, Kistnah District and on Zinc carbonate from Karnul : *Rec. Geol. Surv. Ind.,* XIV, 304—305, (1881).

30. On Iridosmine from the Noa Dihing River, W. Assam, and Platinum from Chutia Nagpore : *Rec. Geol. Surv. Ind.,* XV, 53—55, (1882).

31. On a copper mine near Yongri Hill, Darjiling, Arsenical pyrites from the same neighbourhood and on Kaolin from Darjiling : *Rec. Geol. Surv. Ind.,* XV, 56—58, (1882).

32. Analyses of coal and fire clay from the Makum coal-field, Upper Assam : *Rec. Geol. Surv. Ind.,* XV, 58—63, (1882).

33. On sapphires lately discovered in the N.-W. Himalayas : *Rec. Geol. Surv. Ind.,* XV, 138—141, (1882).

34. New faces on Stilbite from the Western Ghats : *Rec. Geol. Surv. Ind.,* XV, 153—155, (1882).

35. On the Iron ores and Subsidiary materials for the manufacture of Iron in the N.-E. part of the Jabalpur District: *Rec. Geol. Surv. Ind.,* XVI, 94—115, (1883).

36. On Lateritic and other Manganese ores at Gosalpur, Jabalpur District : *Rec. Geol. Surv. Ind.,* XVI, 116—118, (1883).

37. On native lead from Maulmein and Chromite from the Andaman Islands : *Rec. Geol. Surv. Ind.,* XVI, 203—204, (1883).

38. A descriptive catalogue of the collection of minerals in the Geological Museum, Calcutta. 8°. Calcutta, 1883.

39. Popular guide to the Geological collection in the Indian Museum, Calcutta, No. 5, Economic Mineral Products. 8°. Calcutta, 1883.

40. On some of the mineral resources of the Andaman Islands, in the vicinity of Port Blair : *Rec. Geol. Surv. Ind.,* XVII, 79—86, (1884).

41. Notice of a further fiery eruption from the Minbyin mud volcano, Cheduba island, Arrakan : *Rec. Geol. Surv. Ind.,* XVII, 142, (1884).

42. The Volcanoes of Barren Island and Narcondam in the Bay of Bengal : *Mem. Geol. Surv. Ind.,* XXI, 251—286, (1885).

43. On the alleged tendency of the Arrakan Mud Volcanoes to burst into eruption most frequently during the rains : *Rec. Geol. Surv. Ind.,* XVIII, 124—125, (1885).

44. Analyses of Phosphatic nodules and Rock from Mussoorie : *Rec. Geol. Surv. Ind.,* XVIII, 126, (1885).

Mallet, F. R.,—cont.

45. On the mineral hitherto known as Nepaulite: *Rec. Geol. Surv. Ind.*, XVIII, 235—237, (1885).

46. On soundings recently taken off Barren Island and Narcondam by Commander A. Carpenter, R.N., *H.M.I.M.S.* "Investigator": *Rec. Geol. Surv. Ind.*, XX, 46—48, (1887).

47. Note on the "Lalitpur" meteorite: *Rec. Geol. Surv. Ind.*, XX, 153—154, (1887).

Mallet, R. *and* Oldham, T.

1. Notice of some secondary effects of the Earthquake of 18th January, 1869, in Cachar: *Quart. Jour. Geol. Soc.*, XXVIII, 255—270, (1872).

Maltby, F. N.

1. [Mud banks on the Travancore coast]: *Mad. Jour. Lit. Sci.*, XXII, 127—133, (1861).

Manson, E. *and* Batten, J. H.

1. Journal of a visit [by E. Manson] to the Melum and the Oonta Dhoora Pass in Juwahir. Edited by J. H. Batten: *Jour. As. Soc. Beng.*, XI, 1157—1181, (1842).

Marchesetti, C. D.

1. On a Prehistoric Monument of the Western Coast of India: *Jour. Bo. As. Soc.*, XII, 215—218, (1876).

Marcadieu, M.

1. Report on the Ferruginous resources of Dhurmsala, Kangra, Mundee, Kotkhai, Rampoor and Jabal: *Sel. Pub. Corr. Punjab*, II, No. vii, 1—16, (1854).

2. Report on the Kooloo Iron Mines and on a portion of the Mannikurn valley. (Communicated by the Government of India): *Jour. As. Soc. Beng.*, XXIV, 191—202, (1855).

3. Report on the hot mineral spring of Tevah, Kangra District: *Ind. Ann. Med. Sci.*, II, 532—535, (1855).

4. Report on the Thermal Sulphurous source of Lowsah, North East of Noorpoor, Kangra District: *Ind. Ann. Med. Sci.*, II, 536—538, (1855).

5. Report on the determination of Iodine contained in the four saline springs, situated in the Jowala Mokhee valley, and of a spring of the same nature at Kangra: *Sel. Pub. Corr. Punjab*, IV, No. v, 2—14, (1810).

6. Presence of Brome in the Jivah mineral thermal spring and continuation of my report of the 2nd January, 1854; *Sel. Pub. Corr. Punjab*, IV, No. v, 15—19, (1860).

Marcou, J.

1. Geological map of the world. Winterthur, 1861.

2. Carte géologique de la Terre: (with text in 4°). Zurich, 1875.

Marryatt, E. L.

1. Report on a visit to Patent Fuel manufactories (with reference to utilisation of Salt Range Coal): *Prof. Pap. Ind. Eng.*, 3rd series, II, 49—52, (1884).

Marsh, H. C.

1. Description of a trip to the Gilgit Valley, a dependency of the Maharaja of Kashmir: *Jour. As. Soc. Beng.*, XLV, pt. i, 119, (1876).

Marsh, W.

Notes on the occurrence of Gold and other Minerals in Mysore. Bangalore, 1887.

Martin, M.

1. The History, Antiquities, Topography, and Statistics of Eastern India ; comprising the districts of Behar, Shahabad, Bhagulpoor, Goruckpoor, Dinajepoor, Puraniya, Rungpoor and Assam, in relation to their Geology, Mineralogy, Botany, Agriculture, Commerce, Manufactures, Fine Arts, Population, Religion, Education, Statistics, etc., 3 vols., 8°. London, 1838.

Mason, F.

1. The natural productions of Burma, or notes on the fauna, flora and minerals of the Tenasserim provinces and the Burman Empire. 8°. Maulmain, 1850.

Mason, F. *and* Theobald, W.

1. Burma : its people and productions ; or notes on the fauna, flora and minerals of Tenasserim, Pegu, and Burma. Rewritten and enlarged by William Theobald. 2 vols, 8°. Hertford, 1882-1883. [Geology and Mineralogy, vol. I, pp. 1—15.]

Masson, C.

1. Narrative of various journeys in Baloochistan. Afghanistan, and the Punjab, including a residence in those countries from 1826 to 1838. 3 vols., 8°. London, 1842.

2. Narrative of a journey to Kalat, including an account of the insurrection at that place in 1840, and[a memoir on Eastern Baluchistan. 8°. London, 1843.

Masters, J. W.

1. Extract from a memoir of some of the natural productions of the Angami Naga Hills, and other parts of Upper Assam: *Jour. As. Soc. Beng.*, XVII, 57-59, (1848).

Masters, J. W. *and* Prain, D.

1. The Hot springs of the Namba Forest in the Sibsagar district, Upper Assam. Unpublished memoranda by the late J. W. Masters, Esq., with observations by Surgeon D. Prain, *I.M.S.*, Curator of the Herbarium, Royal Botanical Gardens, Calcutta: *Proc. As. Soc. Beng.*, 1887, pp. 201-204.

Mayer, J. E.

1. Quantitative analysis of the spring water of Ramandroog: *Ind. Ann. Med. Sci.*, II, 239-242, (1855).

2. Report on the chemical Examination of several specimens of Salt from the Loonar Lake in the Deccan: *Mad. Jour. Lit. Sci.*, XVII, (new series I), 15-21, (1856).

Mazures, Des, *see* DES MAZURES.

Medlicott, H. B.

1. [Report on Coal in Jummoo.] Coal, No. 13, pp. 13-20, (1859).

2. On the Vindhyan rocks and their associates in Bundelcund : *Mem. Geol. Surv. Ind.*, II, pt. i, 1-95, (1860).

3. On the Sub-Himalayan rocks between the Ganges and the Jumna : *Jour. As. Soc. Beng.*, XXX, 22-31, (1861).

4. Note on the Reh Efflorescence of North-Western India and on the waters of some of the Rivers and Canals : *Jour. Roy. As. Soc.*, XX, 326—344, (1863).

5. On the geological structure and relations of the southern portions of the Himalayan ranges between the rivers Ganges and Ravee : *Mem. Geol. Surv. Ind.*, III, pt. ii, 1—206, (1864). Reviewed in *Geol. Mag.*, 1st decade, II, 310—316, (1865).

6. Note relating to Sivalik Fauna : *Jour. As. Soc. Beng.*, XXXIV, pt. ii, 63—65, (1865) ; *Mem. Geol. Surv. Ind.*, III, pt. ii, 207—209, (1864).

7. The coal of Assam : results of a brief visit to the coal-fields of that province in 1865 : with geological notes on Assam and the hills to the South of it : *Mem. Geol. Surv. Ind.*, IV, pt. iii, 387—442, (1865).

8. On the Alps and Himalayas ; a geological comparison : *Quart. Jour. Geol. Soc.*, XXIII, 322, (1867) ; XXIV, 34—52, (1868).

9. Mode of formation of granite veins at Sungrumpoor, India : *Jour. Geol. Soc Dublin*, I, 87—88, (1867).

10. Notes on the Salt of the Salt Range : *Sel. Rec. Gov. Ind.*, LXIV, 146—155, (1868).

11. [On the action of the Ganges] : *Proc. As. Soc. Beng.*, 1868, p. 232.

12. [On the reported coal of Murree and Kotlee] : *Sel. Rec. Gov. Ind.*, LXIV, (Coal, No. 14), 120—126, (1868).

13. On the prospects of useful coal being found in the Garrow Hills, Bengal. Flscp. Calcutta, 1868 : *Rec. Geol.Surv. Ind.*, I, 11—16,(1868).

14. The boundary of the Vindhyan series to Rájputána : *Rec. Geol. Surv. Ind.*, I, 69—72, (1868).

15. Geological sketch of the Shillong plateau : *Rec. Geol. Surv. Ind.*, II, 10—11, (1869); *Mem. Geol. Surv. Ind.*, VII, pt. i, 151—207, (1869).

16. Memorandum on the wells now being sunk at the European Penitentiary, and at the site for the Central Jail, Hazaribagh : *Rec. Geol. Surv. Ind.*, II, 14—20, (1869).

17. Sketch of the metamorphic rocks of Being : *Rec. Geol. Surv. Ind.*, II, 40—45, (1869).

18. On faults in strata : *Geol. Mag.*, 1st decade, VI, 341—347, (1869); VII, 473—482, (1870).

19. The Mohpani coal-field : *Rec. Geol. Surv. Ind.*, III, 63—70, (1870).

20. Note on the Narbadda coal-basin : *Rec. Geol. Surv. Ind.*, IV, 66—69, (1871).

21. An example of local jointing : *Rec. Geol. Surv. Ind.*, V, 77—79, (1872).

22. Note on exploration for coal in the northern region of the Sátpúra Basin : *Rec. Geol. Surv. Ind.*, V, 109—111, (1872).

23. Note on the Laméta, or infra-trappean, formation of Central India : *Rec. Geol. Surv. Ind.*, V, 115—120, (1872).

Medlicott, H. B.,—cont.

24. Notes on the Satpura coal basin : *Mem. Geol. Surv. Ind.*, X, 133—188, (1873).

25. Sketch of the geology of the North-West Provinces : *Rec. Geol. Surv. Ind.*, VI, 9—17, (1873).

26. Notes on a celt found by Mr. Hacket in the ossiferous deposits of the Narbada valley (pliocene of Falconer) : on the age of the deposits : *Rec. Geol. Surv. Ind.*, VI, 49—54, (1873) ; *Proc. As. Soc. Beng.*, 1873, p. 138.

27. Annual Report of the Geological Survey of India and of the Geological Museum, Calcutta, for the year 1873 : *Rec. Geol. Surv. Ind.*, VII, 1—11, (1874).

28. Note on the habitat in India of the flexible sandstone, or so-called Itacolumyte : *Rec. Geol. Surv. Ind.*, VII, 30—31, (1874).

29. Notes from the Eastern Himalayas : *Rec. Geol. Surv. Ind.*, VII, 53, (1874).

30. Coal in the Garo Hills : *Rec. Geol. Surv. Ind.*, VII, 58—62, (1874).

31. Record of the Khairpur Meteorite of 23rd September, 1873 ; *Jour. As. Soc. Beng.*, XLIII, pt. ii, 33—38, (1874).

32. [Letter on the prospects of an Artesian well at Umballa] : *Prof. Pap. Ind. Eng.*, 2nd series, III, 123—127, (1874).

33. Trials of Raniganj fire-bricks : *Rec. Geol. Surv. Ind.*, VIII, 18—20, (1875).

34. Sketch of the Geology of Scindia's territories : *Rec. Geol. Surv. Ind.*, VIII, 55—59, (1875).

35. The Shahpur coal-field, with notice of coal explorations in the Narbada region : *Rec. Geol. Surv. Ind.*, VIII, 65—86, (1875).

36. Note on the Geology of Nepál : *Rec. Geol. Surv. Ind.*, VIII, 93—101, (1875).

37. Record of the Sitathali Meteorite of 4th March, 1875 : *Proc. As. Soc. Beng.*, 1876, p. 115.

38. Record of the Judesegeri Meteorite of 16th February, 1876 : *Proc. As. Soc. Beng.*, 1876, p. 221.

39. Record of the Nageriá Meteorite of 22nd April, 1876 : *Proc. As. Soc. Beng.*, 1876, p. 222.

40. The retirement of Dr. Oldham : *Rec. Geol. Surv. Ind.*, IX, 27, (1876).

41. Note on the Sub-Himalayan series in the Jamu (Jummoo) Hills : *Rec. Geol. Surv. Ind.*, IX, 49—57, (1876).

42. Remarks on Himalayan glaciation : *Proc. As. Soc. Beng.*, 1877, p. 3.

43. Note on Mr. J. F. Campbell's Remarks on Himálayan Glaciation : *Jour. As. Soc. Beng.*, XLVI, pt. ii, 11, (1877).

44. Annual Report of the Geological Survey of India and of the Geological Museum, Calcutta, for the year 1876 : *Rec. Geol. Surv. Ind.*, X, 1—7, (1877).

45. Observations on underground temperature : *Rec. Geol. Surv. Ind.*, X, 45—48, (1877).

46. Annual Report of the Geological Survey of India, and of the Geological Museum, Calcutta, for the year 1877 : *Rec. Geol. Surv. Ind.*, XI, 1—15, (1878).

Medlicott, H. B.,—cont.

47. Exhibition of the new Geological Map of India : *Proc. As. Soc. Beng.*, 1878, p. 124.

48. Annual Report of the Geological Survey of India and of the Geological Museum, Calcutta, for the year 1878 : *Rec. Geol. Surv. Ind.*, XII, 1—13, (1879).

49. Note on the Mohpani coal-field : *Rec. Geol. Surv. Ind.*, XII, 95—99, (1879).

50. Sketch of the Geology of Rajputana : *Rajputana Gazetteer.* 8°. Calcutta ; Vol. I, pp. 8—19, (1879).

51. Annual Report of the Geological Survey of India and of the Geological Museum, Calcutta, for the year 1879 : *Rec. Geol. Surv. Ind.*, XIII, 1—10, (1880).

52. The Reh Soils of Upper India : *Rec. Geol. Surv. Ind.*, XIII, 273—276, (1880).

53. [Account of some Geological specimens from Afghanistan] : *Proc. As. Soc. Beng.*, 1880, pp. 3—4.

54. Exhibition of a specimen of Rock Salt from the Chakmani territory : *Proc. As. Soc. Beng.*, 1880, p. 123.

55. Annual Report of the Geological Survey of India, and of the Geological Museum, Calcutta, for the year 1880 : *Rec. Geol. Surv. Ind.*, XIV, pp. i—x, (1881).

56. The Nahun-Sivalik unconformity in the N.-W. Himalayas : *Rec. Geol. Surv. Ind.*, XIV, 169—174, (1881).

57. Artesian borings in India : *Rec. Geol. Surv. Ind.*, XIV, 205—238, (1881).

58. Remarks on the Unification of Geological Nomenclature and Cartography : *Rec. Geol. Surv. Ind.*, XIV, 277—279, (1881).

59. Annual Report of the Geological Survey of India, and of the Geological Museum, Calcutta, for the year 1881 · *Rec. Geol. Surv. Ind.*, XV, 1—11, (1882).

60. Note on the supposed discovery of coal on the Kistna : *Rec. Geol. Surv. Ind.*, XV, 207—210, (1882).

61. The Geology of Kumaon and Garhwal : *Gazetteer of North-West Provinces, India.* 8°. Allahabad. Vol. X, pp. 111—168, (1882).

62. Annual Report of the Geological Survey of India, and of the Geological Museum, Calcutta, for the year 1882 : *Rec. Geol. Surv. Ind.*, XVI, 1—9, (1883).

63. Notice of a paper by Captain J. Clibborn, in the Professional Papers of Indian Engineering, on irrigation from wells in the North-West Provinces and Oudh : *Rec. Geol. Surv. Ind.*, XVI, 205—209, (1883) ; *Prof. Pap. Ind. Eng.*, 3rd series, I, 209—211, (1883).

64. Note on Chloromelanite : *Proc. As. Soc. Beng.*, 1883, p. 80.

65. Annual Report of the Geological Survey of India, and of the Geological Museum, Calcutta, for the year 1883 : *Rec. Geol. Surv. Ind.*, XVII, 1—9, (1884).

66. Annual Report of the Geological Survey of India, and of the Geological Museum, Calcutta, for the year 1884 : *Rec. Geol. Surv. Ind.*, XVIII, 1—9, (1885).

Medlicott, H. B.,—cont.

67. Further considerations on Artesian sources in the plains of Upper India : *Rec. Geol. Surv. Ind.*, XVIII, 112—121, (1885).

68. Some observations on percolation as affected by current : *Rec. Geol. Surv. Ind.*, XVIII, 146—147, (1885).

69. Notice of the Pirthalla and Chandpur Meteorites : *Rec. Geol. Surv. Ind.*, XVIII, 148—149, (1885).

70. Preliminary notice of the Bengal Earthquake of 14th July, 1885 : *Rec. Geol. Surv. Ind.*, XVIII, 156—158, (1885).

71. Notice of the Sabetmahet Meteorite : *Rec. Geol. Surv. Ind.*, XVIII, 257—258, (1885).

72. Descriptive list of Exhibits for the Colonial and Indian Exhibition, London, 1886 : 8°.   Calcutta, 1886.

73. Annual Report of the Geological Survey of India, and of the Geological Museum, Calcutta, for the year 1885 : *Rec. Geol. Surv. Ind.*, XIX, 1—9, (1886)

74. Memorandum on the discussion regarding the boulder beds of the Salt Range : *Rec Geol. Surv. Ind.*, XIX, 131—133, (1886).

75. Note on the occurrence of petroleum in India : *Rec. Geol. Surv. Ind.*, XIX, 185—204, (1886).

76. Notice of the Nammianthal meteorite : *Rec. Geol. Surv. Ind.*, XIX, 268, (1886).

77. Annual Report of the Geological Survey of India, and of the Geological Museum, Calcutta, for the year 1886 : *Rec. Geol. Surv. Ind.*, 1—13, (1887).

Medlicott, H. B. *and* Blanford, W. T.

1. A manual of the geology of India, chiefly compiled from the observations of the Geological Survey.   Parts I & II.   8°.   Calcutta, 1879.

Medlicott, J. G.

1. [Note on the Geology of the Narbadda Valley] : *Sel. Rec. Gov. Ind.*, X, 12—29, (1856).

2. Note on the geological structure of parts of Central India : *Jour. As. Soc. Beng.*, XXVIII, 303, (1859).

3. On the Geological Structure of the central portion of the Nerbudda district : *Mem. Geol. Surv. Ind.*, II, pt. i, 101—278, (1860).

Meredith, J.

1. Notes on the topographical features of Assam, and their indications : *Proc. As. Soc. Beng.*, 1869, p. 165.

Merewether, *Sir* W.

1. A Report of the disastrous consequences of the severe earthquake felt on the frontier of Upper Sind on the 24th January, 1852 : *Trans. Bo. Geog. Soc.*, X, 284—286, (1852).

Merewether, *Lieut.*

1. Report on the different localities visited, in 1866, with a view to obtaining stone for the Kurrachee Harbour works : *Prof. Pap. Ind. Eng.*, 1st ser., VI, 130—144, 1869 ; *Trans. Bo. Geog. Soc.*, XVIII, 85—95, (1868).

Meteorites.

1. Account of Meteoric Stones, Masses of Iron and Showers of Dust, Red Snow, and other substances which have fallen from the heavens from the earliest periods down to 1819: *Edin. Phil. Jour.*, I, 221—235, (1819).

Middlemiss, C. S.

1. A fossiliferous series in the Lower Himalayas, Garhwal: *Rec. Geol. Surv. Ind.*, XVIII, 73—77, (1885).

2. Report on the Bengal Earthquake of July 14th, 1885: *Rec. Geol. Surv. Ind.*, XVIII, 200—221, (1885).

3. Physical Geography of West British Garhwal, with notes of a route-traverse through Jaonsar Bawar and Tiri Garhwal: *Rec. Geol. Surv. Ind.*, XX, 26—40, (1887).

4. Crystalline and Metamorphic Rocks of the Lower Himalaya, Garhwal and Kumaon: *Rec. Geol. Surv. Ind.*, XX, 134—143, (1887).

Middleton, G.

1. Analyses made at Colombo of Ceylonese varieties of Ironstone and Lime-stone: *Edin. New Phil. Jour.*, IV, 169, (1828).

Middleton, J.

1. Analysis of cobalt ore found in Western India: *Mem. Chem. Soc. Lond.*, III, 39—41, (1845—1848); *Phil. Mag.*, XXVIII, 352, (1846).

2. Description of the stone common in Agra, Allahabad, Banda, and Mirza-pur: *Sel. Rec. Gov. N.-W. P.*, II, 194—200, (1855); *Ditto*, new series, V, 314—325, (1869).

Middleton, J. E.

1. On the useful ores and earths of Ceylon: *Mad. Jour. Lit. Sci.*, XV, 202—204, (1848).

Miles, R. H.

1. Some remarks upon the country to the South-West of Hoshungabad, and of the Soil, Cultivation, &c., of that part of the valley of the Nerbudda situated between Hoshungabad and the Fort of Mukrai in the lower range of the Kali-bheet Hills: *Jour. As. Soc. Beng.*, III, 61—69, (1834).

Miller.

1. Account of Barren Island: *Cal. Jour. Nat. Hist.*, III, 422—424, (1843).

Mineral.

1. Mineral Wealth of India [notice of articles in the Calcutta *Englishman*]: *Geol. Mag.*, 1st decade, VII, 167—169, (1870).

Mitchell, J.

1. The Mud Bank at Narrikal, near Cochin; its composition as exhibited by the Microscope: *Mad. Jour. Lit. Sci.*, XXII (new series, VI), 264—271, (1861).

Mohun Lall.

1. Account of Kalabagh on the right bank of the Indus: *Jour. As. Soc. Beng.*, VII, p. 25, (1838).

Moir, E. McA.

1. [Review of a report, by E. McArthur Moir, on the Chos of Hoshiarpur] : *Ind. Forester*, X, 271—277, (1884)..

Monckton, E. H. C.

1. [Building stones in Allahabad district] : *Sel. Rec. Gov.*, *N.-W. P.*, II, 187 —190, (1855) ; *Ditto*, new series, V, 302—308, (1869).

Monee Ram.

1. Native account of washing for gold in Assam : *Jour. As. Soc. Beng.*, VII, 621—624, (1838).

Money, W. E.

1. [Building stones in Mirzapur] : *Sel. Rec. Gov.*, *N. W. P.*, II, 183—186, (1855) ; *Ditto*, new series, V, 295—301, (1869).

Montgomerie, *Major* T.

1. Memorandum on the great flood of the river Indus which reached Attok on the 10th August, 1858 : *Jour. As. Soc. Beng.*, XXIX, 128—135, (1860).

2. Narrative Report of the Trans-Himalayan Explorations made in 1867 : *Sel. Rec. Gov. Ind.*, LXXIV, 36—39, (1869).

3. Narrative Report of the Trans-Himalayan Explorations made during 1868 : *Jour. As. Soc. Beng.*, XXXIX, pt. ii, 47—60, (1870).

4. Memorandum on the Results of an exploration of the Namcho, or Tengri Nur Lake, in Great Thibet, made by a native explorer during 1871— 1872 : *Jour. Roy. Geog. Soc.*, XLV, 325—330, (1875). [Fossils.]

Moor.

1. Notices of the Indian Archipelago. 4°. Singapore, 1837.

Moorcroft, W.

1. A Journey to Lake Mánasaróvara in Undés, a province of Little Tibet : *As. Res.*, XII, 375—384, (1816).

2. Notices of the natural productions and agriculture of Cashmere : *Jour. Roy. Geog. Soc.*, II, 253—268, (1832).

Moorcroft, W. *and* Trebeck, G.

1. Travels in the Himalayan provinces of Hindustan and the Punjab ; in Ladakh and Kashmir ; in Peshawar, Kabul, Kunduz and Bokhara, from 1819 to 1825. 2 vols., 8°. London, 1841.

Moore, T. J.

1. Note on the Sivatherium of the Upper Miocene of the Sivalik Hills, and its supposed Zoological relations, as elucidated by Dr. James Murie : *Proc. Liverpool Geol. Soc.*, II, 135, (1872).

Morgan, A.

1. On the Khasi Hill Tribes of North-Eastern Bengal, and on the Geology of the Shillong Plateau : *Proc. Lit. Phil. Soc. Liverpool*, XXX, 115—128, (1876).

Murie, J.

1. On the systematic position of Sivatherium Giganteum : *Brit. Ass. Rep.*, XLI, pt ii, 108—109, 1871. *Geol. Mag.*, 1st decade, VIII, 438—448, 526-527, (1871).

Mornay.

1. Notice of three Trap Dykes in the Burdwan District, and of the effect produced by them on the coal which they pierce: *Cal. Jour. Nat. Hist.*, II, 126—128, (1842).

Morny, S.

1. Qualitative Examination of the Native Copper found on Round Island in the Cheduba Group, south-east of Ramree, and forwarded to the Society by Captain Campbell: *Jour. As. Soc. Beng.*, 904—906, (1843).

Morris, J., *and* Oldham, T.

1. Fossil Flora of the Rajmahal series in the Rajmahal Hills: *Pal. Indica*, series ii, I, 1—52, (1863). [Unfinished, continued in O. FEISTMANTEL, No. 36].

Mosa, P.

1. The Coal-fields of Asia; being an extract translated from Hochstetter's "Asia; its future Railroads and its Coal-fields." 8°. Simla, 1877.

Murchison, *Sir* R. I.

1. Introductory note to Capt. Vicary's Memoir on the Geology of Scinde: *Quart. Jour. Geol. Soc.*, III, 331—333, (1847); WESTERN INDIA, 518—520, (1857).

Murray, A.

1. Notes on Fossil Insects from Nagpur: *Quart. Jour. Geol. Soc.*, XVI, 182—185, (1860).

Murray, J.

1. On the Introduction and use of the natural Mineral waters at Landour: *Cal. Jour. Nat. Hist.*, VIII, 17—40, (1847).

Muzzy.

1. The Geological features of Madura, Trichinopoly, Tanjore, and Poothacotta: *Mad. Jour. Lit. Sci.*, XVII, (new series, I), 90—103, (1857).

N

N. A.

1. Some account of a boring, made in Fort William, for the purpose of procuring a supply of fresh water, with remarks on the nature of that used in Calcutta for domestic consumption: *Glean. Sci.*, I, 102—103, (1829).

2. Details of several Borings made in Calcutta, in search of a spring of Fresh Water: *Glean. Sci.*, I, 167—169, (1829).

Narbadda.

1. Reports and correspondence relative to the Narbadda, the mineral resources of its valley and its navigability: *Sel. Rec. Gov. Ind.*, XIV, 13—145, (1855).

Narra.

1. Reports on the upper portion of the Eastern Narra: *Sel. Rec. Bo. Gov.*, new series, XLV, (1857).

Nelson.

1. Note of an Earthquake and probable subsidence of Land, in the District of Cutch, near the mouth of the Koree. or Eastern branch of the Indus : *Quart. Jour. Geol. Soc.*, II, 103, (1846).

Nelson, J. H.

1. Mineralogy [of the Madura District] : *Madura Manual.* 8°. Madras, 1868. pp. 23—42.

Ness, W.

1. On the Warora Coal-field : *Colliery Guardian*, XXVIII, 745, (1874).

2. The Government Experiments in Iron-making in India : *Colliery Guardian*, XXX, 925, (1875).

3. Report on the Experiment of Iron [at Warora] : *Gazette of India Supplement*, 1876, pp. 480—482.

4. The Warora Coal-field : *Colliery Guardian*, XXXIV, 629—658, (1877).

Neumayr.

1. Die Intertrappean beds im Dekan, und die Laramiegruppe im Westlichen Nordamerika : *Neu. Jahrb. Min. Geol.*, 1884, Band I, pp. 74—76 ; *Rec. Geol. Surv. Ind.*, XVII, 87—88, (1885).

Neville, H.

1. Notes on the geological origin of South-Western Ceylon : *Jour. Ceylon As. Soc.*, 1870—71, pp. 11—20.

Newbold, T. J.

1. A visit to the Gold Mine at Batting Moring, and summit of Mount Ophir, or " Gunong Ledang," in the Malay Peninsula : *Jour. As. Soc. Beng.*, II, 497—502, (1833).

2. Some account of the Territory and Inhabitants of Naning, in the Malayan Peninsula : *Jour. As. Soc. Beng.*, III, 601—616, (1834) ; Moor's *Notices of the Indian Archipelago.* 4°. Singapore, 1837, pp. 246—254.

3. Note on the States of Pérak, Srímenanti, and other States in the Malay Peninsula : *Jour. As. Soc. Beng.*, V, 505—509, (1836).

4. Sketch of the State of Múar, Malay Peninsula : *Jour. As. Soc. Beng.*, V, 561—567, (1836) ; Moor's *Notices, &c.*, Appendix, pp. 73—86.

5. Note on the occurrence of Volcanic Scoria in the Southern Peninsula : *Jour. As. Soc. Beng.*, V, 670—671, (1836).

6. A glance at the Banaganpilly Jaghire, taken while passing through that territory in March, 1836 : *Mad. Jour. Lit. Sci.*, III, 117—122, (1836).

7. Sur les Mines d'étain de Malacca : *Bibl. Univ.*, VIII, 435—437, (1836).

8. On the Régar, or black cotton soil, of India : *Proc. Roy. Soc.*, IV, 53—54, (1838).

9. Sketch of the Malayan Peninsula : *Mad. Jour. Lit. Sci.*, VII, 52—68, (1838).

10. Description of the valley of Sondur : *Mad. Jour. Lit. Sci.*, VIII, 128—152, (1838).

Newbold, T. J.,—cont.

11. Political and Statistical account of the British settlements in the Straits of Malacca, *viz.* Pinang, Malacca, and Singapore, with a history of the Malayan States on the Peninsula of Malacca. *2* vols., 8°. London, 1839.

12. Notice of River Dunes on the Banks of the Hogri and Pennaur : *Mad. Jour. Lit. Sci.*, IX, 307—308, (1839).

13. Some account, Historical, Geographical & Statistical, of the Ceded Districts : *Mad. Jour. Lit. Sci.*, X, 109—132, (1839).

14. The Beryl Mine of Paddioor and geognostic position of this gem in Coimbatoor, Southern India : *Edin. New Phil. Jour.*, XXIX, 241—245, (1840) ; *Mad. Jour Lit. Sci.*, XI, 171—175, (1840).

15. Notes, chiefly Geological, on Southern India, from the Banks of the Tumbuddra to those of the Cauvery : *Mad. Jour. Lit. Sci.*, XI, 126—143, (1840).

16. A cursory note of the Gold Tract in the Kupputgode range. Manganese mines near Wodoorti and Flint excavations in the Southern Mahratta country—the corundum pits of the western part of Mysore and the Diamond mines of Kurnool : *Mad. Jour. Lit. Sci.*, XI, 42—52, (1840)

17. Account of a Carboniferous stratum at Baypore, near Calicut, Malabar Coast : *Mad. Jour. Lit. Sci.*, XI, 239—243, (1840).

18. Note on the geological position of the Laterite in the vicinity of Beder : *Mad. Jour. Lit. Sci.*, XI, 244—245, (1840).

19. Geological Desiderata : *Mad. Jour. Lit. Sci.*, XI, 245—250, (1840).

20. List of minerals for presentation to the society, collected from various parts of the Nizam's Territories, Ceded Districts, Kurnool, the southern Mahratta country, Mysore, &c.: *Mad. Jour. Lit. Sci.*, XII, 16—30, (1840).

21. On the rock-basins in the bed of the Toombuddra, Southern India (Lat. 15° to 16° N.) : *Proc. Geol. Soc.*, 1842, pp. 702—705.

22. Notes, principally geological, on the tract between Bellary and Bijapore : *Jour. As. Soc. Beng.*, XI, 929—940, (1842) ; Western India, 308—316, (1857).

23. Notes, principally geological, from Bijapore to Bellary, *via* Kannighirri : *Jour. As. Soc. Beng.*, XI, 941—955, (1842) ; Western India, 317—327, (1857).

24. Geological specimens offered to the Asiatic Society of Bengal : *Jour. As. Soc. Beng.*, XI, 1131—1135, (1842).

25. On the processes prevailing among the Hindus, and formerly among the Egyptians, of quarrying and polishing granite ; its uses, &c.: with a few remarks, on the tendency of this rock in India to separate by concentric exfoliation : *Jour. Roy. As. Soc.*, VII, 113—128, (1843).

26. On some ancient mounds of scoriaceous ashes in Southern India : *Jour. Roy. As. Soc.*, VII, 129—136, (1843).

27. Mineral resources of Southern India: —No. I, Copper Districts of Ceded Districts, South Mahratta country, and Nellore ; No. II, Magnesite formations ; No. III, Chromate of Iron mines ; Salem District : *Jour. Roy. As. Soc.*, VII, 150—171, (1843).

Newbold, T. J.,—cont.

28. Note on a recent fossil freshwater deposit in Southern India, with a few remarks on the origin and age of the Kanker, and on the supposed decrease of thermal temperature in India : *Jour. As. Soc. Beng.*, XIII, 313—318, (1844) ; *Bibl. Univ.*, LIX, 186—190, (1845) ; *Phil. Mag.*, XXVI, 526—532, (1845).

29. Note on the Osseous Breccia and Deposit in the caves of Billa Soorgum, Lat. 15° 25′, Long. 78° 15′, Southern India : *Jour. As. Soc. Beng.*, XIII, 610—611, (1844).

30. Notes, chiefly geological, across the Peninsula from Masulipatam to Goa, comprising remarks on the origin of the Regur and Laterite ; occurrence of Manganese veins in the latter, and on certain traces of aqueous denudation on the surface of Southern India : *Jour. As. Soc. Beng.*, XIII, 984—1004, (1844) ; WESTERN INDIA, 66—85, (1857).

31. On the Temperature of Springs, Wells, and Rivers of India and Egypt, and of the Sea and Table-Lands within the Tropics : *Phil. Trans.*, 1845, pp. 125—140 ; *Edin. New Phil. Jour.*, XL, 99—115, (1845).

32. On the alpine glacier, iceberg, diluvial and wave translation theories ; with reference to the deposits of Southern India, its furrowed and striated rocks and rock basins : *Jour. As. Soc. Beng.*, XIV, 217—246, (1845).

33 Notes, principally geological, on the South Mahratta country. Falls of Gokauk. Classification of rocks : *Jour As. Soc. Beng.*, XIV, 268—305, (1845) ; WESTERN INDIA, 346—377, (1857).

34. Notes, principally geological, across the Peninsula of Southern India, from Kistapatam, Lat. 14° 17′, at the embouchure of the Coileyroo river, on the Eastern Coast, to Hanawer, Lat. 14° 16′, on the Western coast, comprising a visit to the falls of Gairsuppa : *Jour. As. Soc. Beng.*, XIV, 398—425, (1845).

35. Notes, chiefly geological, across the Peninsula of Southern India, from Madras, Lat. N. 13° 5′, to Goa, Lat. N. 15° 30′, by the Baulpilly Pass and ruins of Bijanugger : *Jour. As. Soc. Beng.*, XIV, 497—521, (1845).

36. Notes, chiefly geological, across the Peninsula from Mangalore in Lat. N. 12° 49′, by the Bisly pass to Madras in Lat. N. 13° 4′ : *Jour. As. Soc. Beng.*, XIV, 641—659, (1845).

37. Notes, chiefly geological, across South India from Pondichery, Lat. N. 10° 56, to Beypoor in Lat. N. 11° 12′, through the great Gap of Palghautcherry : *Jour. As. Soc. Beng.*, XIV, 759—782, (1845).

38. Notes, chiefly geological, on the Coast of Coromandel, from the Pennaur to Pondicherry : *Jour. As. Soc. Beng.*, XV, 204—212, (1846).

39. Notes, chiefly geological, on the Western Coast of South India : *Jour. As. Soc. Beng.*, XV, 224—231, (1846).

40. Notes, chiefly geological, from Seringapatam, by the Hegulla Pass, to Cannanore : *Jour. As. Soc. Beng.*, XV, 315—322, (1846).

41. Notes, chiefly geological, from Koompta on the Western Coast (S. India) by the Devamunni and Nundi Cunnama Passes, easterly to Cumbum,

Newbold, T. J.,—cont.

and thence southerly to Chittoor, comprising a notice of the Diamond and lead excavations of Buswapúr : *Jour. As. Soc. Beng.*, XV, 380—396, (1846).

42. Summary of the geology of Southern India : *Jour. Roy. As. Soc.*, VIII, 138—171, 213—270, 315—318, (1846) ; IX, 1—42, (1848) ; XII, 78—96, (1850).

43. Notes chiefly geological, from Gooty to Hydrabad, South India, comprising a brief notice of the old Diamond Pits at Dhone : *Jour. As. Soc. Beng.*, XVI, 477—486, (1847).

44. On the Thermal springs of Calwa and Mahanandi in the Kurnool Province : *Mad. Jour. Li.. Sci.*, XV, 160—162, (1848).

Nichols, G. J.

1. Note on the Joga neighbourhood and old mines on the Narbadda : *Rec. Geol. Surv. Ind.*, XII, 173—175, (1879).

Nicholson, B. A. R.

1. On the Island of Perim : *Jour. Bo. As. Soc.*, I, 18—25, (1841).

Nicholson, E.

1. The Earth salts of Bellary : *Mad.'Mon. Jour. Med. Sci.*, VI, 1—9, (1872).

Nicolls, W. T.

1. Memorandum to accompany a section and ground-plan of a Fossil Palm Tree discovered at Saugor : *Jour. Bo. As. Soc.*, V, 614—621, (1857).

Nielly, A.

1. Essay on the Geology of Kunkur : *Prof. Pap. Ind. Eng.*, 2nd series, I, 598 —603, (1872).

2. Report on Experiments made on Kunkur mortars and concrete : *Prof. Pap. Ind. Eng.*, 2nd series, II, 115—140, (1873).

3. On Kankar Limes and Cements on the Bari Doab canal : *Prof. Pap. Ind. Eng.*, 2nd series, VI, 378—389, (1877).

Nielley, A., Higham, T., *and* **Brownlow, H. A.**

1. Extracts from reports and letters on Kunkur limes and cements on the Bari Doab canal : *Prof. Pap. Ind. Eng.*, 2nd series, VI, 127—168, (1877).

Nock, J.

1. Report on the road from Sinde, from Subzul to Shikarpoor : *Jour. As. Soc. Beng.*, XII, 59—62, (1843).

Novara.

1. Reise der Oesterreichischen Fregatte "Novara" um die Erde in den Jahren 1857, 1858, 1859, unter den Befehlen des Commodore B. von Wüllerstorf-Urbair. Geologischer Theil. 4° Vienna, 1864 and 1866. Band ii, Abth, I. Geologische Beobachtungen während der Reise der Osterreichischen Fregatte Novara, von F. von Hochstetter.

O

Oakes, R. E.

1. [Letter relating to the discovery of flint implements at Jubbulpore] : *Proc. As. Soc. Beng.*, 1869, pp. 51—53.

Obbard, J.

1. On the Translation of Waves of Water with relation to the great flood of the Indus in 1858 : *Jour. As. Soc. Beng.*, XXIX, 266—282, (1860).

2. On the past and present condition of the River Hooghly : *Sel. Rec. Gov. Ind.*, XLV, 21—37, (1864).

3. Memorandum upon the probable alteration in the channels of the Hooghly from the Diversion of the whole or a portion of the waters of the Damoodha into the Roopnarain : *Sel. Rec. Gov. Ind.*, XLV, 37—42.

Oldfield, J. N.

1. [Iron ores of Heerapoor in Bundelkund] : *Jour. As. Soc. Beng.*, XIII, *Proc.*, p. vii, (1844).

Oldham, C. F.

1. Notes on the lost river of the Indian Desert : *Calcutta Review*, LIX, 1—29, (1874).

Oldham, C. Æ., *and* Boyle, J. A.

1. Geology [of the Nellore District], compiled from notes furnished by Mr. Charles Oldham : *Nellore District Manual.* 8°. Madras, 1873, pp. 40—59.

Oldham, R. D.

1. Note on the Naini Tal landslip : *Rec. Geol. Surv. Ind.*, XIII, 277—282, (1880).

2. Notes on a traverse between Almora and Mussoorie : *Rec. Geol. Surv. Ind.*, 162—164, (1883).

3. Note on the Geology of Jaonsar : *Rec. Geol. Surv. Ind.*, XVI, 193—198, (1883).

4. Report on the Geology of parts of Manipur and the Naga Hills : *Mem. Geol. Surv. Ind.*, XIX, 217—242, (1883).

5. Note on the Earthquake of 31st December, 1881 : *Rec. Geol. Surv. Ind.*, XVII, 47—53, (1884).

6. On the rediscovery of certain localities for fossils in the Siwalik beds : *Rec. Geol. Surv. Ind.*, XVII, 78—79, (1884).

7. Note on the Geology of the Gangasulan Pargana of British Garhwal : *Rec. Geol. Surv. Ind.*, XVII, 161—167, (1884).

8. On the smooth water anchorages of the Travancore coast : *Rec. Geol. Surv. Ind.*, XVII, 190—192, (1884).

9. Rough notes for the construction of a chapter on the History of the Earth : *Jour. As. Soc. Beng.*, LIII, pt. ii, 187—198, (1884) ; *Proc. As. Soc. Beng.*, 1884, pp. 145—147.

10. [On fossil bones from the Jumna alluvium] : *Proc. As. Soc. Beng.*, 1884 pp. 159—161.

11. Note on the probable age of the Mandhali series in the Lower Himalaya : *Rec. Geol. Surv. Ind.*, XVIII, 77—78, (1885).

Oldham, R. D.,—cont.

12. Memorandum on the probability of obtaining water by means of Artesian wells in the plains of Upper India: *Rec. Geol. Surv. Ind.*, XVIII, 110—112, (1885).

13. Notes on the Geology of the Andaman Islands: *Rec. Geol. Surv. Ind.*, XVIII, 135—145, (1885).

14. Memorandum on the correlation of the Indian and Australian coal-measures: *Rec. Geol. Surv. Ind.*, XIX, 39—47, (1886).

15. Memorandum on the prospects of finding coal in Western Rajputana: *Rec. Geol. Surv. Ind.*, XIX, 122—127, (1886).

16. A note on the Olive group of the Salt Range: *Rec. Geol. Surv. Ind.*, XIX, 127—131, (1886).

17. Preliminary note on the Geology of Northern Jessalmer: *Rec. Geol. Surv. Ind.*, XIX, 157—161, (1886).

18. Facetted pebbles from the Salt Range, Punjab: *Geol. Mag.*, 3rd decade, III, 32—35, (1887).

19. On Probable changes in the geography of the Punjáb and its rivers: an Historico-geographical study: *Jour. As. Soc. Beng.*, LV, pt. ii, 322—343, (1887); *Proc. As. Soc. Beng.*, 1886, pp. 171—175.

20. The gneissose rocks of the Himalaya: *Geol. Mag.*, 3rd decade, III, 461—465, (1887).

21. Preliminary sketch of the geology of Simla and Jutogh: *Rec. Geol. Surv. Ind.*, XX, 143—153, (1887).

22. Note on some points in Himalayan geology: *Rec. Geol. Surv. Ind.*, XX, 155—161, (1887).

Oldham, R. D. *and* T.

1. The Cachar earthquake of 10th January, 1869, by the late Thos. Oldham, edited by R. D. Oldham: *Mem. Geol. Surv. Ind.*, XIX, 1—98, (1882).

2. The Thermal springs of India, by the late Thomas Oldham, edited by R. D. Oldham: *Mem. Geol. Surv. Ind.*, XIX, 99—161, (1882).

3. A Catalogue of Indian earthquakes from the earliest time to the end of A.D. 1869, by the late Thomas Oldham: *Mem. Geol. Surv. Ind.*, XIX, 163—215, (1882).

Oldham, T.

1. Remarks on papers and Reports relative to the Discovery of Tin and other ores in the Tenasserim provinces: *Sel. Rec. Beng. Gov.*, VI, 33—44, (1852); BURMA, 366—375, (1882).

2. Report of the Examination of the Districts in the Damoodah valley and Beerbhoom producing Iron ore: *Sel. Rec. Beng. Gov.*, VIII, (1853.)

3. Report on the coal mines of Lakadong in the Jainteah Hills: *Sel. Rec. Beng. Gov.*, XIII, 45—57, (1853).

4. [On the possibility of working Iron ores of Raniganj at a profit]: *Jour. As. Soc. Beng.*, XXII, 486—491, (1853).

5. Memorandum of the results of an examination of gold dust and gold from Shuygween: *Sel. Rec. Beng. Gov.*, XIII, 58, (1853).

6. [Memorandum on coal stated to occur in the Sivok Nuddee, near the river Teesta]: *Jour. As. Soc. Beng.*, XXIII, 201—203, (1854).

Oldham, T.,—cont.

7. On the geological structure of part of the Khasi Hills, with observations on the meteorology and ethnology of that district. 4°. Calcutta, 1854; *Mem. Geol. Surv. Ind.*, I, ii, 99—207, (1858).

8. Notes upon the Geology of the Rajmahal Hills, being the result of Examinations made during the cold season of 1852—1853; *Jour. As. Soc. Beng.*, XXIII, 263—283, (1854).

9. [On the Geological structure of the Sub-Himalayas, south of Darjeeling, the Khasia hills, and of the Rajmahal hills ; and on the age of the coal-bearing rocks of India]: *Jour. As. Soc. Beng.*, XXIII, 617—620, (1854).

10. Geological Report of Ava; H. YULE, *Reports of the mission to Ava.* 4°. Calcutta, 1856, pp. 309—351 (also published separately) ; BURMA, pp. 287—341, (1882).

11. Notes on the Coal-Fields and Tin Stone Deposits of the Tenasserim Provinces : *Sel. Rec. Gov. Ind.*, X, 31—67, (1856); BURMA, pp. 375—406, (1882).

12. Memorandum on Coal found near Thayetmyo on the Irawaddi River : *Sel. Rec. Gov. Ind.*, X, 99—107, (1856); BURMA, pp. 175—182, (1882).

13. Preliminary notice on the coal and iron of Talcheer, in the Tributary Mehals of Cuttack : *Mem. Geol. Surv. Ind.*, I, pt. i, 1—32, (1856).

14. Note on specimens of gold and gold-dust from Shuegween : *Mem. Geol. Surv. Ind.*, I, pt. i, p. 94, 1856.

15. [Account of the results of investigations by the Geological Survey in Central India]: *Jour. As. Soc. Beng.* XXV, 249—255, (1856).

16. General sketch of the districts already visited by the Geological Survey of India: *Brit. Ass. Rep.*, 1857, pt. ii, pp. 85—89; *Edin New Phil. Jour.*, VI, 320—325, (1857).

17. On some additions to the knowledge of the Cretaceous rocks of India : *Jour. As. Soc. Beng.*, XXVII, 112—119, (1858).

18. Annual Report of the Superintendent of the Geological Survey of India, and Director of the Museum of Geology, Calcutta, 1858-59. 8°. Calcutta, 1859.

19. On some Fossil Fish teeth of the genus Ceratodus, from Maledi, south of Nagpur : *Mem. Geol. Surv. Ind.*, I, pt. iii, 295—309, (1859).

20. Report on the Raneegunge coal-field, with special reference to the proposed extension of the line of railway. 8°. Calcutta, 1859 ; *Sel. Rec. Gov. Ind.*, LXIV, 98—110, (1868).

21. Annual Report of the Superintendent of the Geological Survey of India, and Director of the Museum of Geology. Calcutta, 1859—60. 8°. Calcutta, 1860.

22. On the geological relations and probable age of the several systems of rocks in Central India and Bengal : *Mem. Geol. Surv. Ind.*, II, pt. ii, 299—335, (1860).

23. Report on the present state and prospects of the Government Iron Works at Dechouree in Kumaon, May 10th, 1860. Flscp. Calcutta, 1860.

24. [Note on Dr. J. L. Stewart's specimens from the Waziri country] : *Jour. As. Soc. Beng.*, XXIX, 319, (1860).

Oldham, T.,—cont.

25. Annual Report of the Superintendent of the Geological Survey of India and Director of the Museum of Geology, Calcutta, 1860—61. 8°. Calcutta, 1861.

26. [On the age of the Indian Coal-bearing strata]: *Jour. As. Soc. Beng.,* XXX, 177—182, (1861).

27. Annual Report of the Superintendent of the Geological Survey of India and Director of the Museum of Geology, Calcutta, 1861—62. 8°. Calcutta, 1862.

28. Memorandum on the non-existence of "True" slates in India generally, and especially with reference to the slabs of the Kurnool District, Madras Presidency, showing to what purposes they could be applied: *Jour. Roy. As. Soc.,* 1st series, XIX, 31—38, (1862).

29. Annual Report of the Superintendent of the Geological Survey of India and Director of the Museum of Geology, Calcutta, 1862—63. 8°. Calcutta, 1863.

30. Additional remarks on the Geological relations and probable geological age of the several systems of rocks in Central India and Bengal: *Mem. Geol. Surv. India,* III, pt. i, pp. 197—213, (1863).

31. Indian Mineral Statistics, Coal: *Mem. Geol. Surv. Ind.,* III, pt. i, Art. 2, pp. 1—12, (1863).

32. On the occurrence of Rocks of Upper Cretaceous age in Eastern Bengal: *Quart. Jour. Geol. Soc.,* XXIX, 524—526, (1863).

33. Annual Report of the Superintendent of the Geological Survey of India and Director of the Museum of Geology, Calcutta, 1863—64. 8°. Calcutta, 1864.

34. Note on the Fossils in the Society's Collection reputed to be from Spiti: *Jour. As. Soc. Beng.,* XXXIII, 232—237, (1864).

35. Memorandum on the results of a cursory examination of the Salt Range and parts of the districts of Bunnoo and Kohat, with a special view to the mineral resources of those districts: *Sel. Rec. Gov. Ind.,* LXIV, 126—156, (1864).

36. Annual Report of the Superintendent of the Geological Survey of India and Director of the Museum of Geology, Calcutta, 1864-65: 8°. Calcutta, 1865.

37. On stone implements from Madras: *Proc. As. Soc. Beng.,* 1865, p. 206.

38. Annual Report of the Superintendent of the Geological Survey of India and Director of the Museum of Geology, Calcutta, 1865-66. 8°. Calcutta, 1866.

39. Annual Report of the Superintendent of the Geological Survey of India and Director of the Museum of Geology, Calcutta, 1866-67. 8. Calcutta, 1867.

40. Coal Resources and Production of India. Fscp. Calcutta, 1867: *Sel. Rec. Gov. Ind.,* LXIV, 40-73, (1868); *Geol. Mag.,* IV, 264-265, (1867).

41. Annual Report of the Geological Survey of India and of the Museum of Geology, for 1867: *Rec. Geol. Surv. Ind.,* I, 3—9, (1868).

42. Lead in the district of Raipur, Central Provinces: *Rec. Geol. Surv. Ind.,* I, 37, (1868).

Oldham, T.,—cont.

43. Coal in the Eastern Hemisphere: *Rec. Geol. Surv. Ind.,* I, 37—39, (1868).

44. On the agate flake found by Mr. Wynne in the pliocene (?) deposits of the Upper Godavari: *Rec. Geol. Surv. Ind.,* I, 65—69, (1868).

45. Report on the alleged Existence of Coal in the vicinity of Masulipatam: *Gazette of India, Supplement,* 1868, pp. 215—216, 396—397.

46. Annual Report of the Geological Survey of India and of the Museum of Geology for 1868: *Rec. Geol. Surv. Ind.,* II, 25—34, (1869).

47. The coal-field near Chanda, Central Provinces: *Rec. Geol. Surv. Ind.,* II, 94—100, (1869).

48. Lead in the Raipur District, Central Provinces: *Rec. Geol. Surv. Ind.,* II, 101, (1869): *Sel. Rec. Gov. Ind.,* LXXIV, 98, (1869).

49. Meteorites [Khetri fall]: *Rec. Geol. Surv. Ind.,* II, 101, (1869).

50. Mineral Statistics of India, Coal: *Mem. Geol. Surv. Ind.,* VII, 131—150.

51. Notes on the Earthquake of January 10th, 1869: *Proc. As. Soc. Beng.,* 1869, p. 113.

52. Notes on the remains found in a Cromlech at Coorg: *Proc. As. Soc. Beng.,* 1869, 226.

53. Annual Report of the Geological Survey of India and of the Museum of Geology for 1869: *Rec. Geol. Surv. Ind.,* III, 1—10, (1870).

54. The Wardha river coal-fields, Berar and Central Provinces: *Rec. Geol. Surv. Ind.,* III, 45—53, (1870); *Ind. Economist,* I, 186—187, (1870).

55. Explorations for coal in the Wardha River coal-fields: *Ind. Economist,* II, 23—25, (1870).

56. Explorations for coal in the Chanda district: *Ind. Economist,* I, 306, (1870).

57. Annual Report of the Geological Survey of India and of the Museum of Geology for 1870: *Rec. Geol. Surv. Ind.,* IV, 1—14, (1871).

58. On the supposed occurrence of native antimony in the Straits Settlements: *Rec. Geol. Surv. Ind.,* IV, 48, (1871).

59. On the composition of a deposit in the boilers of steam-engines at Raniganj: *Rec. Geol. Surv. Ind.,* IV, 48—49, (1871).

60. Sketch of the Geology of the Central Provinces: *Rec. Geol. Surv. Ind.,* IV, 69—81, (1871); *Central Provinces Gazetteer.* 2nd ed., 8°. Nagpur, I, pp. xxvi—xlvii, (1870).

61. Annual Report of the Geological Survey of India and of the Museum of Geology for 1871: *Rec. Geol. Surv. Ind.* V 1—13, (1872).

29. [Letters regarding the] Discovery of Petroleum near Thayetmyo: *Ind. Economist,* III, 191—193, (1872).

63. Annual Report of the Geological Survey of India and of the Museum of Geology for 1872: *Rec. Geol. Surv. Ind.* VI, 1—7, (18 3).

64. Die Geologische Karte des Salt Range im Pendschab: *Verh. K. K. Geol. Reichs. Wien,* 1873, pp. 168—170.

65. Coal-fields of British India: *Report Rugby School Natural History Society,* 1873, pp. 45—54; *Geol. Mag.,* 2nd decade, I, 269, (1874).

66. Annual Report of the Geological Survey of India and of the Museum of Geology for 1874: *Rec. Geol. Surv. Ind.,* VIII, 1—11, (1875).

Oldham, T.,—cont.

67. Annual Report of the Geological Survey of India and of the Museum of Geology for 1875: *Rec. Geol. Surv. Ind.*, IX, 1—6, (1876).

68. [Letter on the age of the Indian-coal beds] C. B. CLARKE's *Sedimentary Formations of New South Wales*, 4th ed. 8°. Sydney, 1878, p. 57.

Oldham, T. *and* R. D.

1. The Cachar earthquake of 10th January, 1869, by the late Thos. Oldham, edited by R. D. Oldham : *Mem. Geol. Surv. Ind.*, XIX, 1—98, (1882).

2. The thermal springs of India, by the late Thomas Oldham, edited by R. D. Oldham : *Mem. Geol. Surv. Ind.*, XIX, 99—161, (1882).

3. A Catalogue of Indian earthquakes from the earliest time to the end of A.D. 1869, by the late Thomas Oldham : *Mem. Geol. Surv. Ind.*, XIX, 163—215, (1882).

Oldham, T. *and* Mallet, R.

1. Notice of some secondary effects of the Earthquake of 10th January, 1869, in Cachar : *Quart. Jour. Geol. Soc.*, XXVIII, 255—270, (1872)..

Oldham, T. *and* Moris, J.

1. The Fossil flora of the Rajmahal series in the Rajmahal Hills : *Pal. Indica*, series ii, I, 1—52, (1863), [Unfinished, continued in O. Feistmantel, No. 36.]

Oldham, W.

1. On a shower of earth in the Ghazipur District : *Proc. As. Soc. Beng.*, 1868, p. 182.

2. Memorandum on the action of the Ganges : *Proc. As. Soc. Beng.*, 1868, pp. 229—235.

Oliver, E. E.

1. Report on Reh, Swamp, and Drainage of the Western Jumna Canal Districts : *Prof. Pap. Ind. Eng.*, 3rd series I, 63—87, (1883).

2. Note on Coal and Iron in the Punjab : *Sel. Rec. Punjab Gov.*, new series, No. XXI, (1883).

Ommaney, E. L.

1. Note on Patna Boulders : *Jour. As. Soc. Beng.*, XIX, 135—139, (1850).

Orlebar, A. B.

1. Note on the Lake of Lonar : *Trans. Bo. Geog. Soc.*, February, 1839, pp. 35—38.

2. Notes accompanying a collection of geological specimens from Guzerat : *Jour. Bo. As. Soc.*, I, 191—198, (1842).

3. Notes on the Ramghat : *Jour. Bo. As. Soc.*, I, 199—200, (1842).

Orlich, Leopold von.

1. Travels in India, including Sinde and the Punjab. Translated by H. Evans Lloyd. 2 vols., 8°. London, 1845.

O'Riley, E.

1. Notes on the geological formations of Amherst Beach, Tenasserim Provinces : *Cal. Jour. Nat. Hist.*, VIII, 186—189, (1847).

O'Riley, E.,—cont.

2. Rough notes on the Geological and Geographical characteristics of the Tenasserim Provinces : *Jour. Ind. Archip.*, III, 385—411, (1849).

3. Remarks on the Metalliferous Deposits and Mineral Productions of the Tenasserim Provinces : *Jour. Ind. Archip.*, III, 724—743, (1849).

4. The origin of Laterite : *Jour. Ind. Archip.*, IV, 199—200, (1850).

5. Memorandum on Mineral specimens [from Tenasserim] : *Sel. Rec. Beng. Gov.*, VI, 21—29, (1852) ; BURMA, pp. 360—366, (1882).

6. Report on the Henzai Basin ; its streams and the country in its immediate vicinity : *Sel. Rec. Beng. Gov.*, VI, 30—32, (1852).

7. Journal of a Tour east from Toungoo to the Salween River. [Hot springs] : *Sel. Rec. Gov. Ind.*, XX, 49—71, (1856).

8. Journal of a Tour to Karen-ni for the purpose of opening up a Trading Road to the Shan Traders from Mobyay and the adjacent Shan States through that territory, direct to Tungu : *Jour. Roy. Geog. Soc.*, XXXII, 164—216, (1862).

9. Remarks on the "Lake of clear water" in the district of Bassein, British Burma : *Jour. As. Soc. Beng.*, XXXIII, 39—44, (1864).

Ormiston, G. E.

1. Submerged Forest of Bombay Island : *Rec. Geol. Surv. Ind.*, XIV, 320—323, (1881).

Osborne, G.

1. Report of a visit made to the supposed Coal Field at Bidjeegurh (Vijayagadah) : *Jour. As. Soc. Beng.*, 839—847, (1838).

Ouchterlony.

1. [Fossils from Pondicherry] : *Cal. Jour. Nat. Hist.*, II, 112, (1842).

2. Mineralogical Report on a portion of Districts of Nellore, Cuddepah and Guntoor : *Cal. Jour. Nat. Hist.*, II, 283—285, (1842).

Ouseley, J. R.

1. Notice of two beds of Coal discovered near Bara Garahwara in the valley of the Nerbadda : *Jour. As. Soc. Beng.*, IV, 648, (1835).

2. Note on the process of washing for the gold dust and diamonds at Heera Khoond : *Jour. As. Soc. Beng.*, VIII, 1057—1058; (1839).

3. On the course of the river Nerbudda, with a coloured map of the river, from Hoshungabad to Jubbulpoor : *Jour. As. Soc. Beng.*, XIV, 354—356, (1845).

Ouseley, R.

1. [On peat in Pertabgurh] : *Proc. As. Soc. Beng.*, 1865, pp. 85—86.

Owen, J.

1. The Purtabpur stone quarries : *Prof. Pap. Ind. Eng.*, 1st series, II, 81—89, (1865).

2. Method of manufacturing salt as practise_ amongst the Nagas : *Jour. Agri.-Hort. Soc. Ind.*, III, 27—30, (1844).

Owen, R.

1 On the Batracholites, indicative of a small species of Frog (*Rana pusilla*, Ow.) : *Quart. Jour. Geol. Soc.*, III, 224—227, (1847).

Owen, R.,—cont.

2. Note on the Crocodilian Remains accompanying Dr. T. L. Bell's Paper on Kotah: *Quart. Jour. Geol. Soc.*, VIII, 233, (1852); Western India, 307, (1857).

3. Description of the Cranium of a Labyrinthodont reptile (*Brachyops laticeps*) from Mángali, Central India: *Quart. Jour. Geol. Soc.*, X, 473—474, (1854); XI, 37—39, (1855); Western India, 288—290, (1857).

4. Evidences of Theriodonts in Permian Deposits elsewhere than in South Africa: *Quart. Jour. Geol. Soc.*, XXXII, 352—363, (1876).

P

Page, M.

1. Determination of the Alkaline metals in a Lepidote from India: *Chem. News*, XLVIII, 109—110, (1883.)

Palmer, E. C.

1. Report on the result of certain experiments made on the manufacture of Cements and Artificial stones: *Prof. Pap. Ind. Eng.*, 1st series, VII, 253—263, (1870).

Parish, W. H.

1. A report on the Kohistan of the Jullundhur Doab: *Jour. As. Soc. Beng.*, XVII, pt. i, 281, (1848).

2. A Journal of a trip through the Kohistan of the Jullundhur Doab, under taken at the close of the year 1847 and the commencement of 1848, under the orders of the Supreme Government of India, for the purpose of determining the Geological formation of that District: *Jour. As. Soc. Beng.*, XVIII, 360—409, (1849).

Parish, C.

1. Notes of a trip up the Salween: *Jour. As. Soc. Beng.*, XXXIV, pt. ii, 135—146, (1865).

Paske, E. H.

1. [Iron ores of Kangra district]: *Gazette of India Supplement*, 1874, pp. 1482—1487.

Peal, S. E.

1. Note on an extraordinary flood in Upper Assam: *Proc. As. Soc. Beng.*, 1869, pp. 264—265.

2. Extracts from three letters relative to Pot-holes, the Geological structure of Goalpara Hill, and movements of the clouds in Upper Assam: *Proc. As. Soc. Beng.*, 1877, p. 260.

Pearson, A. N.

1. The development of the Mineral resources of India. 8°. Bombay, 1883.

Pellew, F. H.

1. Note on some specimens of Wood and Soil dug out near Baddibati, Hughli district: *Proc. As. Soc. Beng.*, 1873, p. 78.

Pemberton, R. B.

1. Report on the Eastern Frontier of British India; 8°. Calcutta, 1835: *Jour. Roy. Geog. Soc.*, VIII, 391—396, (1838).

2. Report on Bhootan. 8°. Calcutta, 1839.

Pemberton, R. B. *and* Hannay, S. F.

1. Abstract of the Journal of a Route, travelled by Capt. S. F. Hannay, of the 40th Regiment, Native Infantry, from the Capital of Ava to the Amber Mines of the Húkong valley on the South-East frontier of Assam: *Jour. As. Soc. Beng.*, VI, 245—278, (1837).

Pentland, J. B.

1. Description of Fossil Remains of some animals from the N.-E. border of Bengal: *Geol. Trans.*, 2nd series, II, 393—394, (1829); *Proc. Geol. Soc.*, 1834, p. 76; *Glean. Sci.*, I, 186, (1829).

Petersen, T.

1. Skolezit von Poonah: *Neu. Jahrb. Min. Geol.*, 1873, p. 852.

Phayre, Sir A.

1. [Extracts from letters regarding Arracan and Cheduba]: *Cal. Jour. Nat. Hist.*, I, 559—561, (1841).

2. Account of Arracan: *Jour. As. Soc. Beng.*, X, 679—712, (1841).

3. Letter on Stone weapons from Burma: *Proc. As. Soc. Beng.*, 1876, p. 3.

Phillips, F.

1. Discovery of coal deposits in the Lyneeah valley, Scinde: *Jour. Bo. As. Soc.*, VI, 182—184, (1861).

Piddington, H.

1. Analytical Examination of a mineral water from the Athan Hills, Tenasserim Province: *Glean. Sci.*, III, 24—26, (1831).

2. Examination and Analysis of some specimens of Iron Ore from Burdwan: *As. Res.*, XVIII, pt. i, 171—177, (1833); *Glean. Sci.*, I, 295—298, (1830).

3. On the Fertilising Principles of the Inundations of the Hugli: *As. Res.*, XVIII, pt. i, 224—226, (1829).

4. Examination of a Mineral Exudation from Gazni: *Jour. As. Soc. Beng.*, IV, 696—697, (1835); *Bibl. Univ.*, 1836, p. 173—174.

5. A chemical examination of cotton soils from North America, India, the Mauritius and Singapore; with some practical deductions: *Trans. Agri.-Hort. Soc. Ind.*, VI, 198—226, (1839).

6. Examination and Analysis of a soil brought from the island of Chedooba by Captain Halsted of H. M. S. "*Childers:*" *Jour. As. Soc. Beng.*, X, 436—443, (1841).

7. Report on the Soils brought from Chedooba by H. M. S. "*Childers*": *Jour. As. Soc. Beng.*, X, 447—449.

8. Note on the Fossil Jaw sent from Jubbulpore by Dr. Spilsbury: *Jour. As. Soc. Beng.*, X, 620—625, (1841).

9. Museum of Economic Geology of India: *Jour. As. Soc. Beng.*, XI, 322 —326, (1842).

Piddington, H.,—cont.

10. [Sulphur from Karachi]: *Jour. As. Soc. Beng.*, XII, 736—738, (1843).

11. [Argentiferous Galena from Hisato, Chota Nagpore]: *Jour. As. Soc. Beng.*, XII, 738, (1843).

12. Examination of a remarkable red sandstone from the junction of the Diamond Limestone and Sandstone at Nurnoor, in the Kurnool Territory, Southern India. Received for the Museum of Economic Geology for Capt. Newbold, M. N. I., Assistant Commissioner, Kurnool: *Jour. As. Soc. Beng.*, XIII, 336—338, (1844).

13. Chemical Examination of an Aërolite which fell at the village of Mamigaon near Eidulabad, in Khandeish: *Jour. As. Soc. Beng.*, XIII, 884, (1844).

14. Analysis of lignite from Assam: *Jour. As. Soc. Beng.*, XIV, p. lxxxv, (1845).

15. Examination of an Ore of Cerium from Southern India: *Jour. As. Soc. Beng.*, XV, p. lxii, (1846).

16. Report on an ore of Lead and Antimony sent by Lieut.-Colonel Ouseley from Hisato, Chota Nagpore: *Jour. As. Soc. Beng.*, XV, p. lxiv, (1846).

17. Notice of Tremenheerite, a new carbonaceous mineral: *Jour. As. Soc. Beng.*, XVI, 369—371, (1847).

18. On a new kind of Coal, being Volcanic Coal from Arracan: *Jour. As. Soc. Beng.*, XVI, 371—373, (1847).

19. Account of a Volcanic (?) eruption off the Coromandel coast, recorded in the *Asiatic Annual Register*, 1, 1758: *Jour. As. Soc. Beng.*, XVI, 499—500, (1847).

20. Notice of ferruginous spherules imbedded in sandstone from Lullutpore, in Bundelkund: *Jour. As. Soc. Beng.*, XVI, 711—713, (1847).

21. Description and analysis of a new mineral Newboldite, sent from Southern India by Capt. Newbold: *Jour. As. Soc. Beng.*, XVI, 1129—1135, (1847).

22. Examination and analysis of the Ball Coal of the Burdwan Mines: *Jour. As. Soc. Beng.*, XVII, 59—61, (1848); XVIII, 412—413, (1849); XIX, 75—77, (1850).

23. A notice of a remarkable Hot Wind in the Zillah of Purneah: *Jour. As. Soc. Beng.*, XVII, 144—150, (1848).

24. On the great Diamond in the possession of the Nizam: *Jour. As. Soc. Beng.*, XVII, 151—153, (1848).

25. Description and analysis of a large mass of meteoric iron from the Kharackpur Hills, near Monghir: *Jour. As. Soc. Beng.*, XVII, 538-550, (1848); XVIII, 171, (1849).

26. Examination and analysis of an orange-yellow Earth brought from the Sikkim Territory, by Dr. Campbell, Darjeeling, and said to be used there as a cure for Goître: *Jour. As. Soc. Beng.*, XIX, 143—145, (1850).

27. On Calderite, an undescribed Siliceo-Iron and Manganese-Rock, from the district of Burdwan: *Jour. As. Soc. Beng.*, XIX, 145—148, (1850).

28. Examination of the new mineral Haughtonite (a compound of carbonate of lead and sulphate of barytes): *Jour. As. Soc. Beng.*, XIX, 452—454, (1850).

Piddington, H.,—cont.

29. Detailed Report on the copper ores of the Deoghur Mines : *Jour. As. Soc. Beng.*, XX, 1—13, (1851).

30. On a series of Calderite rocks : *Jour. As. Soc. Beng.*. XX, 207—218, (1851).

31. Examination and Analysis of the Shalka Meteorite (Zillah West Burdwan): *Jour. As. Soc. Beng.*, XX, 299—314, (1851).

32. Analysis of Coal and Galena from Assam : *Jour. As. Soc. Beng.*, XX, 366—367, (1851).

33. Index to the Indian Geological, Mineralogical and Palæontological Papers and Analysis in the Journal of the Asiatic Society : *Jour. As. Soc. Beng.*, XX, 409—425, (1851).

34. Second notice on the Argentiferous Ores of Deoghur : *Jour. As. Soc. Beng.*, XXI, 74—76, (1852).

35. On Hircine, a new Resin : *Jour. As. Soc. Beng.*, XXI, 76—79, (1852); XXII, 279—280, (1853).

36. A Table of Analyses of Indian coals : *Jour. As. Soc. Beng.*, XXI, 270—274, (1852).

37. Notice of graphite sent by Capt. Sherwill from Karsiang : *Jour. As. Soc. Beng.*, XXI, 538, (1852).

38. [Analysis and note on a carbonate of lime and iron from the Koràna Hills] : *Jour. As. Soc. Beng.*, XXII, 208—209, (1853).

39. Examination of a sulphuret of copper from the Barragunda Mine, sent by Mr. Mackenzie : *Jour. As. Soc. Beng.*, XXII, 312, (1853).

40. Report on a specimen of Jet coal from the Chawa Nuddee, a tributary to the Teesta, forwarded by Dr. A. Campbell : *Jour. As. So . Beng.*, XXII, 313—314, (1853).

41. On Nepaulite, a new mineral from the neighbourhood of Kathmandoo : *Jour. As. Soc. Beng.*, XXIII, 170—173, (1854).

42. On the quantity of Silt held in suspension by the waters of the Hooghly at Calcutta in each month of the year : *Jour. As. Soc. Beng.*, XXIII, 283—287, (1854); XXV, 151—164, (1856).

43. Examination and Analysis of four specimens of Coal from the neighbourhood of Darjeeling; forwarded by A. Campbell, Esq., Supdt.: *Jour. As. Soc. Beng.*, XXIII, 381—386, (1854).

44. [Account of the peat of Bengal] : *Jour. As. Soc. Beng.*, XXIII, 400—401, (1854).

45. Notes [on the manufacture of iron] : with a catalogue of iron ores, washings and smeltings : *Jour. As. Soc. Beng.*, XXIII, 402—403, (1854).

46. Examination and Analyses of Dr. Campbell's Specimens of Copper Ores obtained in the neighbourhood of Darjeeling : *Jour. As. Soc. Beng.*, XXIII, 477—479, (1854).

47. Examination and analysis of two Specimens of coal from Ava : *Jour. As. Soc. Beng.*, XXIII, 714—717, (1854).

48. Memorandum on the Kunkurs of Burdwan as a flux for smelting the Iron ores, and on some smeltings of iron ores by Mr. Taylor of that district : *Jour. As. Soc. Beng.*, XXIV, 212—214, (1855).

Piddington, H.,—cont.

49. Report on two specimens of Cuttack coal from the Talcheer Mine, forwarded by E. A. Samuell, Esq., Commissioner of Cuttack : *Jour. As. Soc. Beng.*, XXIV, 240—241, (1855).

50. Examination and Analysis of a Coal from Cherrapunji, received from Messrs. Gilmore and McKilligan : *Jour. As. Soc. Beng.*, XXIV, 283—284, (1855).

51. [On the Burdwan paving-stone] : *Jour. As. Soc. Beng.*, XXIV, 703—706, (1855).

52. [Report on iron ores from Cuttack] : *Jour. As. Soc. Beng.*, XXIV, 708—709, (1855).

53. A second series of experiments to ascertain the mean quantity of silt held in suspension by the waters of the Hooghly in various months of the year, as also the quantity carried out to sea, with an appendix on its sectional area and average discharge : *Jour. As. Soc. Beng.*, XXV, 151—164, (1856).

54. Examination of three specimens of Bengal mineral waters [Darjeeling, Jessore, Hazaribagh] : *Jour. As. Soc. Beng.*, XXV, 190—198, (1856).

55. Practical notes on the best mode of obtaining the highest duty from Burdwan coal as compared with English coal : *Jour. As. Soc. Beng.*, XXVI, 254—257, (1857).

Piddington, H. *and* Campbell, A.

1. [Correspondence respecting the discovery of copper at Pushak near Darjiling] : *Jour. As. Soc. Beng.*, XXIII, 206—210, (1854) ; XXIV, 251, 707—708, (1855).

Playfair, G. R.

1. Account of Barren Island : *Sel. Rec. Gov. Ind.*, XXV, 121—123, (1859).

Pollexfen, J. J.

1. Report on the Rajpeepla and adjoining districts, surveyed, during the years 1852 to 1855 : *Sel. Rec. Bo. Gov.*, new series, XXIII, 297—323, (1856).

Postans, J.

1. Report on Upper Sindh and the Eastern portion of Cutchee, with a Memorandum on the Beloochee and other Tribes of Upper Sindh and Cutchee, and a map of part of the country referred to : *Jour. As. Soc. Beng.*, XII, 23—44, (1843).

2. Report on the Muncher Lake, Arrul and Narra Rivers : *Trans. Bo. Geog. Soc.*, June—August, 1839, pp. 122—124 ; *Sel. Rec. Bo. Gov.*, new series, XVII, 389—393, (1855).

Postans, J. *and* Knight, R. C.

1. Reports on the Manchur Lake and Aral and Narra Rivers : *Jour. Roy. As. Soc.*, VIII, 381—389, (1846).

Pottinger, W.

1. On the present state of the river Indus and the route of Alexander the Great : *Jour. Roy. As. Soc.*, I, 199—212, (1834).

Powell, B. H. Baden.

1. Hand-Book of the economic products of the Punjab, with a combined index and glossary of technical vernacular words. 2 vols., 8°. Roorkee and Lahore, 1868 and 1872.

2. The "Chos" of Hoshiarpur: *Sel. Rec. Punjab Gov.*, new series, No. XV, Lahore, 1879.

Powell, G.

1. Extract from a letter regarding the discovery of coal at Nellore : *Mad. Jour. Lit. Sci.*, XVIII, (new series, II), 291—293, (1857).

Prain, D. *and* Masters, J. W.

1. The Hot springs of the Namba Forest in the Sibsagar district, Upper Assam. Unpublished memoranda by the late J. W. Masters, Esq., with observations by Surgeon D. Prain, *I. M. S.*, Curator of the Herbarium, Royal Botanical Gardens, Calcutta : *Proc. As. Soc. Beng.*, 1887, pp. 201—204.

Pratt, Ven'ble Archdeacon J. H.

1. On the Physical difference between a rush of Water like a torrent down a channel and the transmission of Wave down a river—with reference to the Inundation of the Indus, as observed at Attock, in August, 1858 : *Jour. As. Soc. Beng.*, XXIX, 274—282, (1860).

Prinsep, J.

1. Analysis of a Mineral Water : *As. Res.*, XV, Appendix, p. xiv, (1825).

2. On the Rise and Progress of the Lithographic Art in India, with a brief notice of the Native Lithographic stones of that Country : *Glean. Sci.*, I, 54—56, (1829).

3. Examination of the minerals collected by E. Stirling, Esq., at the Turquoise Mines near Nishapur in Persia : *Glean. Sci.*, II, 375—379, (1830).

4. Examination of the water of several Hot springs on the Arrakan coast : *Glean. Sci.*, III, 16—18, (1851).

5. Analyses of the water of the Katkamsandi Hot spring ; of Ghazipur Kankar ; of Iron sand from Raniganj ; of graphite from Ceylon ; of Indian coals : *Glean. Sci.*, III, 278—284, (1831).

6. Examination of a metallic Button, supposed to be Platina, from Ava : *Glean. Sci.*, III, 39—42, (1831).

7. Examination of Minerals from Ava : *Jour. As. Soc. Beng.*, I, 14—17, 305, (1832).

8. Note on the discovery of Platina in Ava : *As. Res.*, XVIII, pt. ii, 279—284, (1833).

9. Note on the Jabalpur Fossil Bones : *Jour. As. Soc. Beng.*, I, 456—458, (1832); II, 583—588, (1833).

10. On the Graphite, or Black-lead, of Ceylon : *Edin. New Phil. Jour.*, 1st series, XIII, 346—347, (1832).

11. Analyses of several Indian, Chinese, and New Holland Coals : *Edin. New Phil. Jour.*, 1st series, XIII, 347—349, (1832).

12. Chemical Analysis of (1) Three specimens of soil from sugarcane fields ; (2) Slaty anthracite from the hills south of Fatehpur in the Hoshunga-

Prinsep, J.,—cont.

bad District, Narbadda; (3) Peat of the Calcutta alluvium dug up from 30 feet below the surface at the Chitpur lock gates; (4) Argentiferous galena from the Borkhampti country on the sources of the Irawaddi river: *Jour. As. Soc. Beng.,* II, 434—438, (1833).

13. Note on the Fossil Bones discovered near Jabalpur: *Jour. As. Soc. Beng.,* II, 583—588, (1833).

14. Note on the coal discovered at Kyuk Phyu in the Arracan District: *Jour. As. Soc. Beng.,* II, 595—597, (1833).

15. On the occurrence of bones of Man in a fossil state: *Jour. As. Soc. Beng.,* II, 632—636, (1833).

16. Illustration of Herodotus' account of the mode of obtaining gold dust in the deserts of Kobi: *Jour. As. Soc. Beng.,* III, 206—207, (1834).

17. Note on the Fossil Bones of the Nerbudda valley, discovered by Dr. G. G. Spilsbury, near Narsinhpur, &c.: *Jour. As. Soc. Beng.,* III, 396—403, (1834).

18. Note on the Fossil Bones of the Jumna River: *Jour. As. Soc. Beng.,* IV, 500—505, (1835).

19. Chemical Analysis: (1) Saltness of the Red Sea; (2) Native Carbonate of Magnesia from South India; (3) Tin from Malacca; (4) Water from the hot springs in Assam; (5) Mineral water from Ava; (6) Sulphuret of Molybdenum: *Jour. As. Soc. Beng.,* IV, 509—514, (1835).

20. Analysis of Copper ore from Nellore; with notice of the Copper Mines at Ajmir and Singhána: *Jour. As. Soc. Beng.,* IV, 574—584, (1835).

21. Analysis of the Nellore copper ores: *Mad. Jour. Lit. Sci.,* IV, 154—159, (1836).

22. Analysis of native Carbonate of Magnesia from South India: *Mad. Jour. Lit. Sci.,* IV, 232—234, (1836).

23. Catalogue of a second collection of fossil bones presented to the Asiatic Society's Museum by Colonel Colvin: *Jour. As. Soc. Beng.,* V, 179—184, (1836).

24. Specimens of the Soil and Salt from the Sámar, or Sambhur, Lake salt-works, collected by Lieut. A. Conolly and analysed by Mr. J. Stephenson: *Jour. As. Soc. Beng.,* V, 798—806, (1836).

25. Report on the Coal discovered in the Tenasserim Provinces by Dr. Helfer: *Jour. As. Soc. Beng.,* VII, 705—706, (1838).

26. Report on ten specimens of Coal from Capt. Burnes: *Jour. As. Soc. Beng.,* VII, 848—854, (1838).

27. Report on the coal beds of Assam, Bengal: *Jour. As. Soc. Beng.,* VII 948—950, (1838).

28. Table of Indian Coal analysed at the Calcutta Assay office, including those published in "*Gleanings in Science,*" September 1831, arranged according to localities, extracted from Report of the Coal Committee: *Jour. As. Soc. Beng.,* VII, 197—199, (1838).

29. Notice of additional fragments of the Sivatherium: *Phil. Mag.,* XII, 40—41, (1838).

Prinsep, J. *and* **Kalikishen, Rajah.**

1. Oriental accounts of the Precious Minerals (translated by Rajah Kalikishen; with remarks by James Prinsep): *Jour. As. Soc. Beng.,* I, 353—363, (1832).

I

Purdon, W.

    1. [On the Korána Hills] : *Jour. As. Soc. Beng.*, XXII : 207—208, (1853).

Q

Quigley, J. H.

    1. Wanderings in the Islands of Interview, (Andaman), Little and Great Coco. 12°. Moulmein, 1850.

R

Rainey, H. J.

    1. What was the Sundarban originally, and when, and wherefore, did it assume its existing state of utter desolation : *Proc. As. Soc. Beng.*, 1868, 264—273.

    2. Note on three maps of the Sundarban : *Proc. As. Soc. Beng.*. 1869, p. 219.

Ranking, J.

    1. Memorandum on the geology of Thayetmyo : *Mad. Jour. Lit. Sci.*, XXI, (new series, V), 55—59, (1859).

Raper, F. V.

    1. Narrative of a Survey for the purpose of discovering the sources of the Ganges : *As. Res.*, XI, 446 - 563, (1810).

Ratton, J. J. L.

    1. The Ultimate source of common salt : *Mad. Jour. Lit. Sci.*, 1879, pp. 135—136.

Ravenshaw, E. C.

    1. Memorandum on the ancient bed of the River Soane and the site of Palibothra : *Jour. As. Soc. Beng.*, XIV, 137—154, (1845).

Ravenshaw, E. J.

    1. On Coal and Gold from the Sub-Himalayas, Moradabad, and Bijnore : *Jour. As. Soc. Beng.*, II, 264—266, (1833).

Raverty, H. G.

    1. An account of the mountain district forming the western boundary of the Lower Derájat commonly called Roh, with notices of the tribes inhabiting it : *Jour. As. Soc. Beng.*, XXVI, 177—206, (1857).

    2. Notes on Káfiristán : *Jour. As Soc. Beng.*, XXVIII, 317—368, [Minerals and Metals, pp. 329—332], (1859).

    3. An account of the Upper and Lower Suwát, and the Kohistán, to the source of the Suwát River: with an account of the tribes inhabiting those valleys : *Jour. As. Soc. Beng.* XXXI, 227—281, (1862).

Reckendorf, Siegmund.

    1. Notes on the Pokree and Dhanpoor copper mines in Gherwal : *Jour. As. Soc. Beng.*, 471—476, (1844).

Reh.

    1. Correspondence relating to the deterioration of Lands from the presence in the soil of Reh : *Sel. Rec. Gov. Ind.* XLII, (1864).

Reh,— cont.

2. Various papers relating to Reh.: *Sel. Rec. Gov., N.-W. P.,* 2nd series, II, 185—213, (1868).

Rennel, J.

1. An account of the Ganges and Barrampooter Rivers : *Phil. Trans.,* 1781, pp. 87—114.

Reyer, E.

1. Zinn in Birma, Siam und Malakka : *Oesterr. Zeitschr. Berg. Hütt.,* 1879, p. 563.

2. Banka und Bilitong, : *Oesterr. Zeitschr. Berg. Hütt.,* XXVII, pp. 384—385, (1879) ; *Nature,* vol. XX, p. 624.

Richardson, D.

1. Journal of a March from Ava to Kendat, on the Khyendwen River, performed in 1831 under the orders of Major H. Burney, the Resident at Ava : *Jour. As. Soc. Beng.,* II, 59—70, (1833).

2. Abstract Journal of an expedition from Moulmein to Ava through the Kareen country, between December, 1836 and June, 1837 : *Jour. As. Soc. Beng.,* VI, 1005—1022, (1837).

3. Journal of a Mission from the Supreme Government of India to the Court of Siam : *Jour. As. Soc. Beng.,* VIII, 1016—1035 ; IX, 219—250, (1840).

4. Notice of Dr. Richardson's recent journey into Northern Laos : *Glean. Sci.,* II, 211—216, (1830).

Richardson, D. *and* **Blundell, E. A.**

1. An account of some of t he petty states lying north of the Tenasserim Provinces, drawn up from the journals and reports of Dr. D. Richardson [by E. A. Blundell] : *Jour. As. Soc. Beng.,* V, 601—625, 688—707, (1836).

Ricketts, H.

1. [On the copper mines of Dhalbhum and Singhbhum] : *Jour. As. Soc. Beng.,* XXIII, 396—398, (1854).

2. Report on the districts of Singbhoom : *Sel. Rec. Beng. Gov.,* XVI, 63—116, (1854).

Rigby, C. P.

1. Account of a collection of geological specimens for presentation to the Museum of the Bombay Asiatic Society : *Trans. Bo. Geog. Soc.,* VII, 128—129, (1846).

2. Report on Turan Mull, in the Satpura mountains : *Trans. Bo. Geog. Soc.,* IX, 1—9, (1849).

3. On the Satpoora mountains : *Trans. Bo. Geog. Soc.,* IX, 69—98, (1849).

Righy.

1. Memorandum on the usual Building Materials of the district of Cuttack, forwarded to the Museum of Economic Geology, with a set of specimens : *Jour. As. Soc. Beng.,* XI, 836—838, (1842).

Rink, H.

1. Letter to Dr. McClelland on the geological structure of the Nicobars: *Cal. Jour. Nat. Hist.*, VII, 207—213, (1847).

2. Die Nikobarischen Inseln, Eine geographische Skizze, mit specieller Berücksichtigung der Geognosie. 8°. Copenhagen, 1847; translated in *Sel. Rec. Gov. Ind.*, LXXVII, 109—154, (1870).

Roberts, W.

1. Report on the natural products of British Singrowlee, Zilla Mirzapore: *Sel. Rec. Gov., N.-W. P.*, III, 481—484; *Ditto*, new series, III, 146—152, (1867).

Robertson, A. C.

1. Memoranda on the Mud Volcanoes in the district of Luss: *Jour. Bo. As. Soc.*, III, 8—20, (1851).

Robertson, J.

1. Topographical remarks regarding Afghanistan, made during the advance and residence of H. M.'s 13th Light Infantry between 1st April, 1839, and 31st March, 1840: *Cal. Jour. Nat. Hist.*, II, 327—342, (1842).

Robinson, S. H.

1. Section at Dhobah Colliery, Burdwan District: *Cal. Jour. Nat. Hist.*, III, 418—419, (1843).

Robinson, W.

1. A descriptive account of Assam, with a sketch of the local geography. 8° Calcutta, 1841.

Rogers, Alex.

1. Discovery of nummulitic limestone *in situ* at Turkeysur in Surat: *Jour. As. Soc. Bo.*, VI, 164, (1861).

2. A few remarks on the Geology of the country surrounding the Gulf of Cambay in Western India: *Quart. Jour. Geol. Soc.*, XXVI, 118—124, (1870).

Rogers, M. W.

1. Memorandum on the Earthquake of the 31st December, 1881, and the great sea-waves resulting therefrom, as shown on the diagrams of the tidal observatories in the Bay of Bengal: *Proc. As. Soc. Beng.*, 1883, pp. 63—66; *Rep. Surveyor General Ind.* 1881—1882, pp. 71—73.

Rogers, T. E.

1. Correspondence regarding the coal-beds in Namsang, Naga Hills: *Jour. As. Soc. Beng.*, XVII, pt. i, 489—491, (1848).

Romanis, R.

1. Note on borings for coal at Engsein: *Rec. Geol. Surv. Ind.*, XV, 138, (1882).

2. On the outcrop of coal in the Myanoung division of the Henzada district: *Rec. Geol. Surv. Ind.*, XV, 178—181, (1882).

3. Report on the Yenanchaung oil wells. Flscp. Rangoon, 1884.

4. Report on the oil wells and coal of the Thayetmyo district, British Burma: *Rec. Geol. Surv. Ind.*, XVIII, 149—151, (1885).

Romanis, R.,—cont.
5. The gold-fields of Burma: *Chem. News*, LIV, 278—279, (1886).
6. Notes on Upper Burma: *Trans. Edin. Geol. Soc.*, V, pt. ii, 306, (1887) *Scottish Geog. Mag.*, II, 431, (1886).

Rose, R.
1. Account of the process of making Iron at Amdeah, near Sambhalpúr: *Glean. Sci.*, III, 330, (1831).

Ross, D.
1. Notice of some tin ore from the coast of Tenasserim: *Glean. Sci.*, I, 143, (1829).

Ross, W. A.
1. On Jeypoorite, a Sulph-antimonial arsenide of cobalt: *Proc. Roy. Soc. Lond.*, XXI, 292—297, (1873).
2. Meerschalumite [indurated China clay] from Simla: *Ind. Economist*, I, 65 —92, (1869).

Roth, J. R.
1. [Remarks on a Hippopotamus, supposed Palæotherium, skull from the Narbudda alluvium]: *Jour. As. Soc. Beng.*, X, 627—628, (1841).

Row, J.
1. Geological remarks during the march from Benares (old road), *via* Hazareebaugh, Bankoora, and Burdwan to Barrackpoor: *Jour. As. Soc. Beng.*, XIII, 862—866, (1844).

Rowlatt, E. A.
1. Report of an expedition into the Mishmee Hills to the north-east of Sudyah: *Jour. As. Soc. Beng.*, XIV, 477—496, (1845).

Royle, J. F.
1. Extracts from Explanatory address on the exhibition of his collections in Natural History at the meeting of the Asiatic Society on the 7th March [Geology of Dehra-Dun and Raniganj coal fossils]: *Jour. As. Soc. Beng.*, I, 96—100, (1832).
2. Illustrations of the Botany and other branches of Natural History of the Himaláyan Mountains and of the Flora of Kashmir: *Jour. As. Soc. Beng.*, III, 530, (1834); *Jour. Roy. Geog. Soc.*, 361—365, (1835).
3. Illustrations of the Botany and other branches of the Natural History of the Himalayan Mountains and of the Flora of Cashmere: 2 vols., 4°. London, 1839.
4. Essay on the productive resources of India. 8°. London, 1840.
5. Geological features of the Himalayan Mountains: *Mad. Jour. Lit. Sci.*, XI, 323—343, (1840).
6. Report on a specimen of Iron ore from Malwan in the Southern Concan: *Jour. Bo. As. Soc.*, I, 139—142, (1842).
7. On the Tin Mines of the Tenasserim Province: *Proc. Geol. Soc. Lond.*, IV, 165—167, (1843).

Royle, J. F.,—cont.

8. Observations on the Graphite or Plumbago of Kumaon and of Travancore (Communicated by the Government of India) : *Jour. As. Soc. Beng.,* XXIV, 203—206, (1855) ; *Mad. Jour. Lit. Sci.,* XVII, (new series, I), 257—261, (1857).

Rundall, C. *and* Boyle, J. A.

1. Soils [of the Nellore District], compiled from notes furnished by Mr. Charles Rundall: *Nellore District Manual.* 8°. Madras ; 1873 ; pp. 40 —59.

Rundall, F. H.

1. The River Systems of South India : *Proc. Roy. Geog. Soc.,* new series, VIII, 681—698, (1886).

Ryan, Jer.

1. Gold-mining in India ; its past and present. 8°. London, 1880.

S

Saint Hilaire, Isidore Geoffroy.

1. Sur le nouveau genre *Sivatherium,* trouvé fossile au bas du versant méridional de l'Hymalaia, considéré comme devant se rapporter au genre Cameleopardalis : *Comptes Rendus,* IV, 53—60, (1837).

2. Du Sivatherium de l'Hymalaïa comme offrant un cas analogue de terrain, et de degré d' organisation à l' éléphant mammouth, etc.: *Comptes Rendus,* IV, 77—82, (1837).

3. Nouvelles considérations sur le Sivatherium et sur les conséquences relatives à la physiologie zoologique qui se déduisent de la comparaison de ce fossile avec d'autres animaux de la même époque et avec des espèces de l'époque actuelle : *Comptes Rendus,* IV, 113—120, (1837).

Saise, W.

1. The Kurhurballee Coal-Field, with some remarks on Indian Coals: *Trans. N. Eng. Inst. Min. Mech. Eng.,* XXX, (1880).

Salter, J. W. *and* Blanford, H. F.

1. Palæontology of Niti in the Northern Himalaya : being descriptions and figures of the palæozoic secondary fossils, collected by Col. Richard Strachey. 8°. Calcutta, 1865.

Samuells, E. A.

1. [Account of a visit to the coal-fields of Talcheer] : *Jour. As. Soc. Beng.,* XXIV, 248—250, (1855).

Sankey, R. H.

1. On the Geology of some parts of Central India : *Quart. Jour. Geol. Soc.,* X, 55—56, (1854).

2. On the Limestones of Mysore : *Prof. Pap. Ind. Eng.,* 2nd series, III' 26—38, (1874).

Satlaj.

Table showing the breadth of the river and the rate of its current at different stages, from Harríke Pattan to its junction with the Indus at Mithankot: *Jour. As. Soc. Beng.*, V, 814—815, (1836).

Saunders, W.

On Hydraulic Cements: *Glean. Sci.*, III, 54—57, (1831).

Saxton, G. H.

1. [Coal from north of Sumbhulpore]: *Jour. As. Soc. Beng.*, XXIV, 185—186, 354, (1856).

2. [Letter regarding the fall of an Aerolite at Nidigullam, Vizagapatam district, on 23rd Jan., 1870]: *Proc. As. Soc. Beng.*, 1870, 64.

Schindler, A. H.

1. The Turquoise Mines of Nishapur, Khorassan: *Rec. Geol. Surv. Ind.*, 132—142, (1884).

Schlagintweit, Adolphe.

1. Report on the Progress of the Magnetic Survey, and of the Researches connected with it from November 1855 to April 1856, [Rocks of the Cuddapah District]: *Jour. As. Soc. Beng.*, XXVI, 97—110, 122—132, (1857); *Mad. Jour. Lit. Sci.*, XX, (new series, IV,) 32—341, (1859).

2. Official reports on the [last journeys and] death of Adolphe Schlagintweit: *Mad. Jour. Lit. Sci.*, XX, (new series, IV,) 304—332, (1859).

Schlagintweit, Hermann.

1. Report on the Progress of the Magnetic Survey and of the Researches connected with it in Sikkim, the Khasia Hills, and Assam, April to December 1855: *Jour. As. Soc. Beng.*, XXV 1—30, (1856).

2. Report on the Proceedings of the Magnetic Survey, from January to May, 1856: *Jour. As. Soc. Beng.*, XXV, 554—569, (1856): *Mad. Jour. Lit. Sci.* XX, (new series, IV), 341—356, (1859).

3. Assam, das mittlere Stromgebiet des Brahmaputra: *Abhandl. Naturhist. Ges. Nurnberg*, III, 30—41, (1864).

4. Reisen in Indien und Hochasien: 4 vols., 4°. Jena, 1869—1880.

5. Untersuchungen uber die Salzseen in Westlichen Tíbet und in Turkistán. I Theil. Rúpchu und Pangkong: das gebíet der Salzseen im Westlichen Tibet: *Abhandl. K. Baier Akad. Wiss.*, XI, 101—174, (1873).

6. Ueber nephrit nebst Jadeit und Saussurit im Kûnlûn Gebirge: *Sitz. K. Baier Akad. Wiss.*, III, 227—267, (1873).

Schlagintweit, Robert von.

1. Report on the Progress of the Magnetic Survey and of the Researches connected with it, from November 1855 to April 1856: *Jour. As. Soc. Beng.*, XXVI, 54—62, (1857).

2. Report on the Proceedings of the Officers engaged in the Magnetic Survey of India: *Jour. As. Soc., Beng.*, XXVI, 208—216, (1857).

3. Ueber Erosionsformen der Indischen Flüsse.: *Zeits. Allg. Erdkund.*, III, 428—431, (1857).

4. Enumeration of the Hot Springs in India and High Asia: *Jour. As. Soc. Beng.*, XXXIII, 49—56, (1864),

Schlagintweit, Robert von,—cont.

5. Physicalish-Geographische Schilderung von Hoch Asien, *Petermann Mitth.* 1865, pp. 361—377: *Jour. As. Soc. Beng.,* XXXV, pt. ii, 51—72, (1866); *Proc. As. Soc. Beng.,* 1866, pp. 21—22.

6. [Remarks on nephrite from Turkestan]: '*Proc. Boston Soc. Nat. Hist.,* XII, 138, (1869).

Schlagintweit, Adolphe *and* Robert von.

1. Report upon the Progress of the Magnetic Survey of India, and of the Researches connected with it in the Himalaya Mountains from April to October 1855: *Jour. As. Soc. Beng.,* XXV, 105—133, (1856).

Schlagintweit, Herman *and* Robert von.

1. On erosion of rivers in India: *Brit. Ass. Rep.,* 1857, pt. ii, pp. 90—91.

2. An account of a Journey across the chains of the Kuenluen, from Ladak to Khotan: *Jour. As. Soc. Beng.,* XXVI, 111—121, (1857).

Schlagintweit, Adolphe, Herman *and* Robert von.

1. Results of a Scientific Mission to India and High Asia, undertaken between the years 1854 and 1858, by order of the Court of Directors of the Honourable East India Company. With 4 folio atlases of panoramas, views and maps. 4 vols., 4°. London and Leipzig, 1861—66.

Schwager, C.

1. Fossile Foraminiferen von Kar Nikobar: NOVARA, 187—268, (1864).

Schwalbe, B.

1. Magneteisenstein von Landu in Bengalen: *Vierteljahrs Naturf. Ges. Zurich,* VII, 258—263, (1862); *Zeits Gesammt. Naturw Halle,* XX, 199—201, (1862).

Schwartz, C. R. Ritter von.

1. Iron ores in India: *Jour. Iron Steel Inst.,* 1886, pp. 225—229; from *Oesterr. Zeits. Berg. Hüttenw* XXXIII, 715—717, 734—736, 751—754, 771—774, (1886).

Scott, A. J.

1. Analyses of Indian ores of Manganese and some Scottish Zeolites: *Edin. New Phil. Jour.,* 1st series, LIII, 277—285, (1852).

Scott, D.

1. On the banks of the Tísta and Súbúk Rivers: *Geol. Trans.,* 2nd series, I, 137—140, (1824).

2. Notice accompanying specimens from the neighbourhood of the Garo village of Robagiri: *Geol. Trans.,* 2nd series, I, 167, (1824).

Scott, G. F.

1..Report on the Choi Coal Exploration: *Rec. Geol. Surv. Ind.,* XVII, 73—78, (1884).

Severn, H. A.

1. [Gold Mining in India]: *Jour. Soc. Arts,* XXX, 592, (1882).

Shastree, Ball G.
1. Note on a specimen of iron ore from the vicinity of Malwan; *Jour. Bo. As. Soc.*, I, 436, (1844).

Shaw, R. B.
1. Visits to High Tartary, Yârkand and Kâshgar (formerly Chinese Tartary), and return journey over the Karakorum pass. 8°. London, 1871.

Shephard, C. L.
1. Mineralogical notices [Syhedrite] : *Am. Jour. Sci.*, 2nd series, XL, 110—112, (1865).

Sheppard, G. F.
1. [Account of Carnelians in the Rajpipla hills] : *Bombay Gazetteer*, 8°. Bombay, 1880, VI, 198—207.

Sherwill, J. L.
1. Journal of a Trip undertaken to explore the Glaciers of the Kanchunjingah Group in the Sikkim Himálaya, in November 1861 : *Jour. As. Soc. Beng.*, XXXI, 457—479, (1862).

Sherwill, S. R.
1. The Kurrukpur Hills : *Jour. As. Soc. Beng.*, XXI, 195—206, (1852).

Sherwill, W. S.
1. Note on a curious Sandstone formation at Sasseram, Zillah Shahabad : *Jour. As. Soc. Beng.*, XIV, 496—498, (1845).
2. [Corundum mines in Singrowlee] : *Jour. As. Soc. Beng.*, XIV, p. xv, (1845).
3. Note on the geological features of Zillah Behar : *Jour. As. Soc. Beng.*, XV, 55—60, (1846).
4. Geological notes on Zillah Shahabad or Arrah : *Jour. As. Soc. Beng.*, XVI, 279—285, (1847).
5. A sketch of the Behar Mica Mines : *Jour. As. Soc. Beng.*, XX, 295—298, (1851).
6. Notes upon a Tour through the Rájmahal Hills : *Jour. As. Soc. Beng.*, XX, 544—606, (1851).
7. Geological map of Bengal. Calcutta, 1852.
8. Notes upon a Tour in the Sikkim Himálaya Mountains, undertaken for the purpose of ascertaining the Geological Formation of Kunchinjinga and of the perpetually snow covered peaks in its vicinity : *Jour. As. Soc. Beng.*, XXII, 540—570, 611—638, (1853).
9. Report on the Rivers of Bengal : *Sel. Rec. Beng. Gov.*, XXIX, 1—18, (1858).

Shortt, J.
1. Report on the Medical Topography of the South Western Political Districts, embracing the partly unexplored country between Latitude 20° 48′ and 22° 12″ N. and Longitude 83° 15′ and 84° 10″ E. : *Sel. Rec. Mad. Gov.*, XIV, (1855).

Sibley, G.
1. Diagram of floods on the Jumna River during the years 1861—65 : *Prof. Pap. Ind. Eng* 1st series, II, 242, (1873).

Sibold, E. A.

1. On the retrogression of level in canals : *Prof. Pap. Ind. Eng.*, 2nd series, II, 141—144, (1873).

2. On the alluvion and diluvion of the Punjab rivers : *Prof. Pap. Ind. Eng.*, 2nd series, VIII, 235—242, (1879).

Silliman, B., *Jr.*

1. On a granular Albite associated with Corundum and on the Indianite of Bournon : *Am. Jour. Sci.*, 2nd series, VIII, 389—391, (1849).

2. On "Indianite" of Count Bournon and on the American mineral which has been distributed under the same name : *Am. Assoc. Proc.*, II, 131—134, (1849).

Sipöcz, L.

1. Kaliglimmer Aus Ostindien : *Mineral Mitth.*, 1873, pp. 31—32.

Smart, R. B.

1. Description of country surveyed in District Raepore, Season, 1867—1868 : *Sel. Rec. Gov. Ind.*, LXXIV, 95—99, (1869).

Smith, A.

1. On Earthquakes at Chittagong : *Proc. As. Soc. Beng.*, 1866, p. 39.

Smith, David.

1. Report on the coal and iron districts of Bengal : 8°. (Calcutta?), 1856 : *Sel. Rec. Gov. Ind.*, LXIV, 74—89, (1868).

2. Report on the Singrowlee and Kurhurbalee coal-fields : 8°. Calcutta, 1857 : *Sel. Rec. Gov. Ind.*, LXIV, 90—98, (1868).

Smith, E.

1. Notes on specimens of the Kankar Formation and on fossil bones collected in the Jumna : *Jour. As. Soc. Beng.*, II, 622—636, (1833).

Smith, G.

1. Description and plan of the Natron Lake of Loonar, with an analysis of the Salt [by J. E. Mayer] : *Mad. Jour. Lit. Sci.*, XVII, (new series, I), 1—21, (1856).

Smith, R. Baird.

1. On the crystalline structure of the Trap Dykes in the Sienite of Amboor : with an enquiry into the causes to which this peculiarity of certain Igneous rocks is due : *Mad. Jour. Lit. Sci.*, IX, 287—307, (1834).

2. Notes illustrative of the geology of Southern India : *Mad. Jour. Lit. Sci.*, XI, 315—323, (1840) ; *Cal. Jour. Nat. Hist.*, I, 324—343, (1841).

3. On the structure of the Delta of the Ganges as exhibited by the Boring-operations in Fort William, A. D. 1836—1840 : *Proc. Geol. Soc.*, IV, 4—6, (1842) ; *Cal. Jour. Nat. Hist.*, I, 324—343, (1841).

4. Memorandum on the organization of a museum of Economic geology for the North-Western Provinces of British India, to be established at Agra : *Jour As. Soc. Beng.*, X, 779—794, (1841).

5. On the Geological Relations of Artesian Wells : *Cal. Jour. Nat. Hist.*, II, 20—31, (1842).

Smith, R. Baird,—cont.

6. Economic Geology: *Cal. Jour Nat. Hist.*, II, 16—31, (1842).

7. Notes on the recent earthquakes on the North-Western Frontier: *Jour. As. Soc. Beng.*, XI, 242—254, (1842).

8. Memoir on Indian Earthquakes: *Jour. As. Soc. Beng.*, XII, 2c8—292, 1029—1056, (1843); XIII, 964—983, (1844); XIV, 964—983, (1845).

9. Remarks on Earthquakes in British India: *Edin. New Phil. Jour.*, 1st series, XXXIV, 107—109, (1843).

10. Fictitious vegetable impressions on Sandstone Rocks: *Cal. Jour. Nat. Hist.*, III, 22—24, (1843).

11. Register of Indian and Asiatic Earthquakes for the year 1843: *Jour. As. Soc. Beng.*, XIV, 604—622, (1845).

Smith, A. Pye.

1. Report on the manufacture of Ransome's artificial stone in Bombay, 1869: *Prof. Pap. Ind. Eng.*, 1st series, VII, 263—270, (1878).

Smithe, J. D.

1. On the Gravels and Boulders of the Punjab: *Quart. Jour. Geol. Soc.*, XVII, 163, (1861).

Smyth, R. Brough.

1. Report on the gold mines of the south-eastern portion of the Wynaad and the Carcoor ghat. Flscp. Madras, 1880.

Sowerby, J. D. C.

1. [Letter acknowledging receipt of fossils from the Himalayas, with list of the same]: *Jour. As. Soc. Beng.*, I, 248—249, (1832).

Sowerby, W.

1. Observations on the deposits of Iron ore in the lower hills and Bhabur, at Loha Bhurbur and Dechouree, made on the 28th and 29th August 1855: *Sel. Rec. Gov. Ind.*, VIII, Supplement, 44—49, (1855).

2. Progress Reports of the survey of the Iron Deposits in the Bhabur from 22nd September to 15th December 1855: *Sel. Rec. Gov. Ind.*, XVII, 7—18, 25—31, (1856).

3. General Summary of the Preliminary Survey of the Iron deposits in the lower hills of Kumaon and Garhwal from the Golah River to the Ganges: *Sel. Rec. Gov. Ind.*, XVII, 33—47, 49—51, (1856).

4. List of specimens of Bhabar Iron ores, with Remarks: *Sel. Rec. Gov. Ind.*, XVII, 78—85, (1856).

5. Report of the survey of the mineral deposits in Kumaon, and on the iron smelting operations experimentally conducted at Dechouree: *Sel. Rec. Gov. Ind.*, XVIII, 1—56, 87—107, (1856).

6. Some account of the recently discovered deposits of iron-ore at the foot of the Himalayas, in Kumaon, Northern India: *Proc. Inst. Civ. Eng.*, XVI, 82—84, (1857).

7. Report on the Government Iron Works at Kumaon, with remarks on the Iron deposits of the Himalayas: *Sel. Rec. Gov. Ind.*, XXVI, 1—93, (1859).

8. [Mineral resources of India]: *Jour. Soc. Arts*, XXX, 593—594, (1882).

Sowerby, W.

1. Memorandum on the geological action on the South coast of Kattywar and in the Runn of Kutch: *Trans. Bo. Geog. Soc.*, XVIII, 96—104, (1868).

2. Tides and their action geologically and geographically considered: *Trans. Bo. Geog. Soc.*, XIX, pp. lxxx—lxxxiv, 121—143, (1874).

Spilsbury, G. G.

1. On two hot springs in the valley of the Nerbudda: *Trans. Med. Phys. Soc. Calcutta*, III, 450—451, (1827).

2. Announcement of despatch of a fossil bone from Jabalpur: *Jour. As. Soc. Beng.*, II, 45, 263, (1833).

3. Fossil shells from Jabalpur: *Jour. As. Soc. Beng.*, II, 205, (1833).

4. Geological section across the valley of the Nerbudda from Tendukhéri to Bittoul: *Jour. As. Soc. Beng.*, III, 388—395, (1834).

5. Notice of new Sites of Fossil deposits in the Nerbudda valley: *Jour. As. Soc. Beng.*, VI, 91, 487—489, (1837); VII, 91, (1838).

6. [Notice of discovery of coal near [Jubbulpore]: *Jour. As. Soc. Beng.*, VIII, 530, (1839).

7. On fifteen varieties of Fossil Shells found in the Saugor and Nerbudda territories: *Jour. As. Soc. Beng.*, VIII, 708—711, (1839).

8. Notes on various Fossil Sites on the Nurbudda; illustrated by specimens and drawings: *Jour. As. Soc. Beng.*, VIII, 950—952, (1839).

9. Notes of a march from Brimhan Ghat on the Nerbudda to Umurkuntuk, the source of that River: *Jour. As. Soc. Beng.*, IX, 889—903, (1840).

10. Notes on Fossil Discoveries in the valley of the Nerbudda: *Jour. As. Soc. Beng.*, X, 626—627, (1841).

11. Notes on the Nerbudda Fossils: *Jour. As. Soc. Beng.*, XIII, 765, (1844).

Sprenger, A.

1. As-Soyúti's work on Earthquakes called Kashf as Salsalah 'an wasf Azzalzalah, *i.e.*, removing the noise from the description of Earthquakes (or clearing up the description of Earthquakes). Translated from the Arabic: *Jour. As. Soc. Beng.*, XII, 741—748, (1843).

Spry, H. H.

1. Note on Indian Saline deposits: *Jour. As. Soc. Beng.*, I, 503, (1832).

2. Note on the fossil Palms and Shells lately discovered on the Table-land of Ságar in Central India: *Jour. As. Soc. Beng.*, I, 561, (1832); II, 376—377, 549—550, 639—641, (1833).

3. A three weeks' sail in search of health. Province of Arracan—Khyok-Phyoo—its Harbour, Productions, Capabilities, Geological features; visit to an Active Volcano: *Jour. As. Soc. Beng.*, X, 138—147, (1841).

4. Extract from letter regarding the Cap Island [Ramree] Coal: *Cal. Jour Nat. Hist.*, II, 1.7—119, (1842).

Steel, E. H.

1. [On an Earthquake in Upper Assam, 11th April 1870]: *Proc. As. Soc. Beng.*, 1870, p. 222.

Steel, E. H.,—cont.

2. [Letter on Jade celts found in Assam]: *Proc. As. Soc. Beng.*, 1870, p. 267.

Steinbach.

1. The Punjab; being a brief account of the country of the Sikhs; its extent, history, commerce, productions, government, manufactures, laws, religion, &c. 8°. London (2nd edition), 1846.

Stephenson, J.

1. On the Saline Nature of the soil of Ghazipoor, and Manufacture of common salt, as practised by the Natives of the villages of Tuttulapoor, Ratouly, Sahory, Chilar and Becompoor: *Jour. As. Soc. Beng.*, III, 36—39, (1834).

2. On the Efflorescence of Khári Nún, or Sulphate of Soda, as found native in the soil of Tirhút and Sarun in the province of Behar: *Jour. As. Soc. Beng.*, III, 188—189, (1834).

3. Note on Vegetable Impressions in Agates: *Jour. As. Soc. Beng.*, IV, 507—508, (1835).

4. Specimens of the Soil and Salt from the Sámar or Sambhur Lake Salt Works, collected by Lieut. Arthur Conolly: *Jour. As. Soc. Beng.*, V, 802—804, (1836).

Stevenson, J.

1. On the Manufacture of Saltpetre as practised by the Natives of Tirhút: *Jour. As. Soc. Beng.*, II, 23—26, (1833).

2. Notice of a Native Sulphate of Alumina from the Aluminous Rocks of Nipal: *Jour. As. Soc. Beng.*, II, 321, (1833).

3. Notice of a Native Sulphate of Iron from the hills of Behar, and used by Native Dyers of Patna: *Jour. As. Soc. Beng.*, II, 321-322, (1833).

4. Note on the Salàjit of Nipal [alum and ammonites]: *Jour. As. Soc. Beng.*, II, 605, (1833).

5. Note on the Pea Stalactite of Tibet: *Jour. As. Soc. Beng.*, IV, 520, (1835).

Stevenson, J. F.

1. Account of a visit to the hot springs of Pai in the Tavoy district: *Jour. As. Soc. Beng.*, XXXII, 383—386, (1863).

Stewart, J.

1. Geological notes on the strata between Malwa and Guzerat: *Trans. Lit. Soc. Bombay*: III, 538—542, (1821)

2. Account of a bed of native sub-carbonate of Soda found in Malwa: *Trans. Lit. Soc. Bombay*, III, 53—54, (1823).

Stewart, J. L.

1. Extracts from letters on the geology of the Waziri country: *Jour. As. Soc. Beng.*, XXIX, 314—318, (1860).

Stiffe, A. W.

1. On the Mud-Craters and Geological Structure of the Mekran Coast: *Quart. Jour. Geol. Soc.*, XXX, 50—53, (1874).

Stirling, Andrew.

1. An account, Geographical, Statistical and Historical, of Orissa Proper or Cuttack : *As. Res.*, XV, 163—338, (1825).

Stirling, W.

1. A visit to the Falls of Sansadurra : *Trans. Bo. Geog. Soc.*, 1841—1844, pp. 5—7.

2. Notice of granite protruding through the trap in the bed of the River Nerbudda at Mundleysir: *Trans. Bo. Geog. Soc.*, 1841—1844, pp. 7—8.

Stoehr, E.

1. [Letter on Geology and Copper of Singhbum] : *Neu. Jahrb. Min. Geol.*, 1857, pp. 47—50.

2. Einige Bemerkungen über den District Singhbum in Bengalen : *Vierteljahrs. Naturf. ges. Zurich* ; V, 329—362, (1860); *Rec. Geol. Surv. Ind.*, III, 86—92, (1870).

3. Das Kupfererz-Vorkommen in Singhbum, Provinz der Sudwestgrenze von Bengalen: *Neu. Jahrb. Min., Geol.*, 1864, pp. 129—159; *Rec. Geol. Surv. Ind.*, III, 86—92, (1870).

Stoliczka, F.

1. Cretaceous Fauna of Southern India : *Pal. Indica*, series i, iii, v, vi & viii.

> Vol. I. The Ammonitidæ, pp. 41—216, (1861—1865).
> „   II. The Gastropoda (1867—1868).
> „ III. The Pelecypoda (1870—1871).
> „ IV. The Brachiopoda, Ciliopoda, Echinodermata, Corals, &c., (1872—1873).

2. On the character of the Cephalopoda of the South Indian Cretaceous rocks : *Quart. Jour. Geol. Soc.*, XXI, 407—412, (1865).

3. Geological sections across the Himalayan mountains from Wangtu Bridge on the river Sutlej to Sungdo on the Indus, with an account of the formations in Spiti, accompanied by a revision of all known fossils from that district : *Mem. Geol. Surv. Ind*, V, pt. i, 1—154, (1865).

4. Geologische Schreiben aus Simla : *Sitz. K. K. Akad. Wien*, 1865, pp. 379—382.

5. Geologische Schreiben aus Kaschmir : *Sitz. K. K. Akad. Wien*, 1866, pp. 104—123.

6. Summary of geological observations during a visit to the Provinces of Rupshu, Karnag, South Ladak, Zanskar, Suroo and Dras of Western Tibet, in 1865 : *Mem Geol. Surv. Ind.*, V, pt. iii, 337—354, (1866).

7. On Jurassic Deposits in the North West Himalaya: *Quart. Jour. Geol. Soc.*, XXIV, 506—509, (1865).

8. Additional observations regarding the cephalopodous fauna of South Indian Cretaceous deposits : *Rec. Geol. Surv. Ind.*, I, 32—37, (1868).

9. General results obtained from an examination of the gastropodous fauna of the South Indian Cretaceous deposits : *Rec. Geol. Surv. Ind*, I, 55—59, (1868).

Stoliczka, F.,—cont.

10. Die Andaman-Insel, Assam, u. s. w.: *Verh. K. K. Geol. Reichs. Wien.*, 1868, pp. 157—159.

11. Osteological notes on *Oxyglossus Pusillus (Rana pusilla*, Owen), from the tertiary frog beds in the Island of Bombay: *Mem. Geol Surv. Ind.,* VI, 387—394, (1869).

12. Note on *Pangshura tecta* and other species of Chelonia, from the newer tertiary deposits of the Nerbudda Valley: *Rec. Geol. Surv. Ind.,* II, 36—39, (1869).

13. Reisen in Hinter-Indien, auf die Nikobaren und Andamanen: *Verh. K. K. Geol. Reichs. Wien ;* 1870, pp. 23—28.

14. Note on the Kjokkenmoddings of the Andaman Islands: *Proc. As. Soc. Beng.*, 1870, pp. 13—23.

15. Tertiary crabs from Sind and Kach: *Pal. Indica,* series vii, xiv, Vol. I, (1871).

16. A brief account of the geological structure of the hill-ranges between the Indus Valley in Ladák and Sháh-i-dula on the frontier of the Yárkand territory: *Rec. Geol. Surv. Ind.,* VII, 12—15, (1874); T. D. Forsyth's *Report, &c.,* pp. 460—462; Yárkand, No. 1, pp. 15—18, (1878).

17. Geological notes on the route traversed by the Yárkand Embassy from Sháh-i-dula to Yárkand and Káshgar: *Rec. Geol. Surv. Ind.,* VII, 49—51, (1874); *Quart. Jour. Geol. Soc.,* XXX, 571—575, (1874); T. D. Forsyth's *Report, &c.,* pp. 462—464; Yárkand, No. 1, pp. 21—23.

18. Note regarding the occurrence of jade in the Karakásh Valley, on the southern borders of Turkistán: *Rec. Geol. Surv. Ind.,* VII, 51—53, (1874); *Quart. Jour. Geol. Soc.,* XXX, 568—570, (1874); T. D. Forsyth's *Report, &c.,* pp. 464—466; Yárkand, No. 1, pp. 18—20.

19. Geological observations made on a visit to Chaderkul, Thian Shan range: *Rec. Geol. Surv. Ind.,* VII, 81—85, (1874); *Quart. Jour. Geol Soc.,* XXX, 574—580, (1874); T. D. Forsyth's *Report, &c.,* pp. 466-470; Yárkand, No. 1, pp. 24—29.

20. Note on the Pamir: *Rec. Geol. Surv. Ind.,* VII, 86, (1874).

21. The Altum-Artash considered from a geological point of view: *Rec. Geol. Surv. Ind.,* VIII, 13—16, (1875); T.D. Forsyth's *Report, &c.,* pp. 470—473; Yárkand, No. 1, pp. 30—33.

Stoliczka, F., *and* **Blanford, H. F.**

1. Catalogue of the Specimens of Meteoric Stones and Meteoric Irons in the Museum of the Asiatic Society of Bengal, Calcutta, corrected up to January, 1866: *Jour. As. Soc. Beng.,* XXXV, pt. ii, 43—45, (1866).

Stoliczka, F., *and* **Blanford, W. T.**

1. Scientific results of the second Yarkand Mission, based upon the collections and notes of the late Ferdinand Stoliczka: Geology, by W. T. Blanford. 4°. Calcutta, 1878; Yarkand, No. 1.

Stone.

1. Stone and Marble quarries at Mirzapore: *Cal. Jour. Nat. Hist.,* I, 429—430, (1861).

2. Quarries in the North-West Provinces: *Sel. Rec. Gov. N.-W. P.,* II, 146—149, 178—200, (1855); *Ditto,* new series, V, 276—313, (1869).

Storer, F. H. *and* **Warren, C. M.**

1. Examination of Naphtha obtained from Rangoon Petroleum : *Mem. Am. Acad.*, new series, IX, 208—216, (1867).

Strachey, H.

1. Narrative of a Journey to Cho Lagan (Rakas Tal), Cho Mapan (Manassarówar) and the valley of Pruang in Gnari, Hundés, in September and October, 1846 : *Jour. As. Soc. Beng.*, XVII, pt. ii, 98—120, 125—182, 327—551, (1818).

2. Physical geography of Western Thibet : *Jour. Roy. Geog. Soc.*, XXIII, 1—68, (1853).

Strachey, R.

1. A Description of the Glaciers of the Pindur and Kuphilee rivers in the Kumaon Himalaya : *Jour. As. Soc. Beng.*, XVI, 794—812, (1847) ; *Edin. New Phil. Jour.*, XLIV, 108—123, (1847).

2. Note on the Motion of the Glacier of the Pindur in Kumaon : *Jour. As. Soc. Beng.*, XVII, pt. ii, 203—205, (1848) ; *Edin. New Phil. Jour.*, XLVI, 258—262, (1849).

3. Notice of a Trip to the Niti Pass : *Jour. As. Soc. Beng.*, XIX, 79-82, (1850).

4. Notice of Lieut. Strachey's Scientific Enquiries in Kumaon : *Jour. As. Soc. Beng.*, XIX, 239—242, (1850).

5. On the geology of part of the Himalaya mountains and Thibet : *Quart. Jour. Geol. Soc.*, VII, 292—310, (1851) ; *Brit. Ass. Rep.*, 1857, pt. ii, p. 69.

6. On the Physical geography of the Provinces of Kumaon and Garhwal in the Himalayan Mountains and the adjoining parts of Thibet : *Jour. Roy. Geog. Soc.*, XXI, 57—85, (1851).

7. On the physical geology of the Himalayas : *Quart. Jour. Geol. Soc.*, X, 249—253, (1854).

8. Report on a project for the establishment of iron works in the Kumaon Bhabur. 8°. Agra, (1856).

Strong, Dr.

1. Report on the progress of the boring experiment [in Fort William] : *Jour. As. Soc. Beng.*, I, 250—251, 298, 473, 561, (1832) .

Strover, G. A.

1. Memorandum on the metals and minerals of Upper Burma : *Gazette of India Supplement*, (1873) ; *Ind. Economist*, V, 13—14, (1873) ; *Geol. Mag.*, 1st decade X, 356—361, (1873).

Summers, A.

1. Statement of the wrought Agates, Cornelians, &c., together with the varied processes of preparation, and value of the Trade at Cambay : *Sel. Rec. Bo. Gov.*, new series, IV, 26—36, (1854).

2. An account of the agate and cornelian trade of Combay : *Jour. Bo. As. Soc.*, III, 318—327, (1851).

Swiney, J. D.

1. Letter regarding the discovery of Flint implements. near Jubbulpore, in the Narbadda alluvium : *Jour. Bo. As. Soc.*, VIII, p. xvii, (1864).

Swynnerton, C.

1. On a Celt of the Palæolithic type, found at Thandiani, Punjab, September 10th, 1880, by Charles Massy Swynnerton: *Proc. As. Soc. Beng.*, 1880, p. 175.

Sykes, W. H.

1. [Geology of] a portion of the Dekhan: *Geol. Trans.*, 2nd series, IV, 409—432, (1835); *Proc. Geol. Soc.*, 1834, pp. 417—419; *Mad. Jour. Lit. Sci.*, VI, 344—374, (1837); Western India, 39—115, (1857); *Bombay Gazetteer.* 8° Bombay, 1885, XV, 9—12.

2. A Notice respecting some Fossils collected in Cutch by Captain Walter Smee, of the Bombay Army: *Geol. Trans.*, 2nd series, V, 715—719, (1840); *Proc. Geol. Soc.*, 1840, pp. 715—720; Western India, 460—466, (1857).

3. On a Fossil Fish from the Table-land of the Deccan, in the peninsula of India, with a description of the specimens by Sir P. de M. Egerton: *Quart. Jour. Geol. Soc.*, VII, 272—273, (1851); *Jour. Bo. As. Soc.*, V, 146—148, (1857); Western India, 301—302, (1857).

Symes.

1. An account of an Embassy to the kingdom of Ava, sent by the Governor General of India in 1795. 4°. London, 1800. 2nd ed., 3 vols., 8°, 4° atlas, London, 1800.

2. A brief account of the Religion and Civil Institutions of the Burmans; and a description of the Kingdom of Assam, to which is added an account of the Petroleum Wells, in the Burma Dominions, extracted from a Journal from Rangoon up the River Erawaddy to Amaraporah, the present capital of the Burmah Empire. 8°. Calcutta, 1826.

T

Tapp, H.

1. [Mineral Wealth of India] : *Jour. Soc. Arts*, XXX, 592, (1882).

Tagore, *Raja Sir* Sourindro Mohun.

1. Mani-Málá : or a Treatise on gems. 2 parts, 8°. Calcutta, 1879—1881.

Tavernier, Jean Baptiste.

1. Collections of travels through Turkey into Persia, and the East Indies, giving an account of the present state of those countries. As also a full relation of the five years' wars between Aureng-Zebe and his brothers in their father's life-time, about the succession. And a voyage made by the Great Mogul (Aureng-Zebe) with his army from Delhi to Lahor, from Lahor to Bember, and from thence to the kingdom of Kachemire, by the Mogols called the Paradise of the Indies. Together with a relation of the kingdom of Japan and Tunkin, and of their particular manners and trade. To which is added a new description of the Grand Seignor's seraglio, and also of all the kingdoms that encompass the Euxine and Caspian Seas. 4°. London, 1684.

Tavoy.

1. Remarks on the geology of Tavoy: *Cal. Jour. Nat. Hist.*, II, 359—367, (1842).

Taylor, R.

1. On changes of Madras Coast : *Proc. As. Soc. Beng.* 1866, pp. 51—52.

Taylor, R. F.

1. Iron-smelting and the Napier Foundry, Madras : *Ind. Economist,* VI, 131—133, (1875).

Taylor, R. Meadows.

1. Sketch of the geology of the district of Shorapur or Soorpoor in the Dekhan : *Jour. Roy. Geol. Soc. Dublin,* X, 24—33, (1862).

Taylor, T. G.

1. Memoranda regarding a boring executed on the Beach at Madras : *Mad. Jour. Lit. Sci.,* XIV, 183—187, (1847).

Taylor, T. M.

1. Note on the progress of the Boring in Fort William : *Jour. As. Soc. Beng.,* V, 374—375, (1836).

2. Progress report of the Boring Experiment in Fort William : *Jour. As. Soc. Beng.,* VI, 234—237, (1837) ; *Mad. Jour. Lit. Sci.,* VII, 470—472, (1838).

Temple, R. C.

1. Notes on the Formation of the Country passed through by the 2nd Column, Tal Chotiali Field Force, during its march from Kala Abdullah Khan in the Khójak Pass to Lugárí Bárkhán, spring of 1879 : *Jour. As. Soc. Beng.,* XLVIII, pt. ii, 103—109, (1879).

2. An account of the country traversed by the second column of the Thal Chotiali Field Force in the spring of 1879, [Geological note by H. B. Medlicott and O. Feistmantel] : *Jour. Roy. Geog. Soc.,* XLIX, 230—319, (1879).

Tenasserim.

1. List of specimens of rocks from the Tenasserim Archipelago : *Jour. As.* II, 157, (1833).

Tennent, *Sir* J. Emerson.

1. Ceylon : an account of the island, physical, historical, and topographical, with notices of its natural history, antiquities, and productions. 2 vols. 8°. London, 5th ed., 1860.

2. Sketches of the natural history of Ceylon, with narratives and anecdotes illustrative of the habits and instincts of the mammalia, birds, reptiles, fishes, insects, &c., including a monograph of the elephant and a description of the modes of capturing and training it : 8°. London, 1861.

Thackeray, E. T.

1. Report on the Artesian Boring at Umballa : *Prof. Pap. Ind. Eng.,* 2nd series, III, 117—123, (1874).

Theobald, W.

1. Notes on the geology of the Punjab Salt Range : *Jour. As. Soc. Beng.,* XXIII, 651—678, (1854).

Theobald, W.,—cont.

2. On the tertiary and alluvial deposits of the central portion of the Nerbudda valley : *Mem. Geol. Surv. Ind.*, II, 279—335, (1860).

3. Note on stone celts from Bundelkhund and chert implements from the Andamans : *Jour. As. Soc. Beng.*, XXXI, 323—328, (1862).

4. Notes of a trip from Simla to the Spiti valley and Chomoriri (Tshomoriri) lake during the months of July, August and September, 1861 ; *Jour. As. Soc. Beng.*, XXXI, 480—532, (1862).

5. Notes on the occurrence of Celts in British Burma : *Proc. As. Soc. Beng.*, 1865, pp. 126—127.

6. Notes on the stone implements of Burma : *Proc. As. Soc. Beng.*, 1869, pp. 181—186.

7. On the beds containing silicified wood in Eastern Prome, British Burma : *Rec. Geol. Surv. Ind.*, II, 79—86, (1869) ; BURMA, pp. 182—194, (1882).

8. Note on some Agate Beads from North-Western India : *Proc. As. Soc. Beng.*, 1869, p. 253.

9. Remarks on a stone implement from Burmah : *Proc. As. Soc. Beng.*, 1870, p. 220.

10. On the alluvial deposits of the Irawadi, more particularly as contrasted with those of the Ganges : *Rec. Geol. Sur. Ind.*, III, 17—27, (1870) ; BURMA, pp. 194—209, (1882).

11. On petroleum in British Burma, &c. : *Rec. Geol. Surv. Ind.*, III, 72—73, (1870) ; BURMA, pp. 209—210, (1882).

12. The axial group in Western Prome, British Burma : *Rec. Geol. Surv. Ind.*, IV, 33—44, (1871) ; BURMA, pp. 210—226, (1882).

13. A few additional remarks on the Axial group of Western Prome : *Rec. Geol. Surv. Ind.*, V, 79—82, (1872) ; BURMA, pp. 226—230, (1882).

14. Note on the value of the evidence afforded by raised oyster banks on the coasts of India, in estimating the amount of elevation indicated thereby : *Rec. Geol. Surv. Ind.*, V, 111—112, (1872).

15. A brief notice of some recently discovered petroleum localities in Pegu *Rec. Geol. Surv. Ind.*, V, 120—122, (1872) ; BURMA, pp. 230—232, (1882).

16. On the Geology of Pegu : *Mem. Geol. Surv. Ind.*, X, 189—359, (1873) ; BURMA, pp. 1—172, (1882).

17. Notes of a celt found by Mr. Hacket in the ossiferous deposits of the Narbada Valley, (pliocene of Falconer). On the associated shells : *Rec. Geol. Surv. Ind.*, VI, 54—57, (1873).

18. On the salt springs of Pegu : *Rec. Geol. Surv. Ind.*, VI, 67—73, (1873) ; BURMA, pp. 232—238, (1882).

19. Stray notes on the metalliferous resources of British Burma : *Rec. Geol. Surv. Ind.*, VI, 90—95, (1873) ; BURMA, 406—413, (1882).

20. On the former extension of glaciers in the Kángra District : *Rec. Geol. Surv. Ind.*, VII, 86—98, (1874).

21. Remarks on certain considerations adduced by Falconer in support of the antiquity of the human race in India : *Rec. Geol. Surv. Ind.*, VII, 142—145, (1874).

Theobald, W.,—cont.

22. Letter forwarding two perforated stone Implements found at Kharakpur, in the Monghyr District : *Proc. As. Soc. Beng.* 1875, p. 102.

23. Remarks on Mr. Campbell's paper on Himalayan Glaciation, in the Journal, No. 1, part ii, 1877 : *Proc. As. Soc. Beng.*, 1877, p. 137.

24. Description of a new emydine from the upper tertiaries of the Northern Punjáb : *Rec. Geol. Surv. Ind.*, X, 43—45, (1877).

25. On the occurrence of erratics in the Potwár, and the deductions that must be drawn therefrom : *Rec. Geol. Surv. Ind.*, X, 140—143, (1877).

26. Remarks, explanatory and critical, on some statements in Mr. Wynne's paper on the tertiaries of the North-West Panjáb, in Records (vol. X, part iii) : *Rec. Geol. Surv. Ind.*, X, 223—235, (1877).

27. [Geology of] Guzerat : BOMBAY, pp. 1—14, (1878).

28. On a marginal bone of an undescribed Tortoise from the upper Sivaliks, near Nila, in the Potwar, Punjab : *Rec. Geol. Surv. Ind.*, XII, 186—187, (1879).

29. On the Kamaon lakes : *Rec. Geol. Surv. Ind.*, XIII, 161—175, (1880).

30. On the discovery of a celt of palæolithic type in the Punjab : *Rec. Geol. Surv. Ind.*, XIII, 176, (1880).

31. On some Pleistocene deposits of the N. Punjab and the evidence they afford of an extreme climate during a portion of that period : *Rec. Geol. Surv. Ind.*, XIII, 221—243, (1880).

32. Geology and Economic mineralogy [of Lower Burma] ; *Brit. Burma Gazetteer*, I, 32—67, (1880).

33. The Siwalik group in the Sub-Himalayan Region : *Rec. Geol. Surv. Ind.*, XIV, 66—125, (1881).

Theobald, W., Blanford, W. T. and H. F.

1. On the Geological Structure and Relations of the Talcheer coal-field in the district of Cuttack : *Mem. Geol. Surv. Ind.*, I, pt. i, 33—89, (1856).

Theobald, and Mason, F.

1. Burma, its people and productions : or Notes on the fauna, flora, and minerals of Tenasserim, Pegu, and Burma. Rewritten and enlarged by William Theobald. 2 vols., 8°. Hertford, 1882—1883. [Geology and Mineralogy, vol. I, pp. 1—15.]

Thomas, E. C. G.

1. On the iron ores and the manufacture of Iron and Steel of Coimbatore : *Appendix, Rep. Gov. Mus. Madras*, 1856, pp. 17 [separately paged].

Thomas, R. H.

1. Miscellaneous information connected with the Province of Cutch : *Sel. Rec. Bo. Gov.*, new series, XV, (1855).

Thomason, C. S.

1. On the drainage and irrigation of the Terai : *Prof. Pap. Ind. Eng.* 1st series, I, 422—436, (1863).

Thomson, Murray.

1. [Note on an approximate method of] Kunkur analysis : *Prof. Pap. Ind. Eng.*, 2nd series, I, 491—496, (1872).

2. Note on coal from the Sivalik Hills : *Prof. Pap. Ind. Eng.*, 2nd series, VII, 117—118, (1878).

Thomson, R. D.

1. Chemical analysis of an Indian specimen of Mesolite : *Edin. New Phil. Jour.*, 1st series, XVII, 186—189, (1834).

2. Sketch of the geology of Bombay Islands : *Rec. Gen. Sci.*, I, 291—304, 330—341, (1835) ; *Mad. Jour. Lit. Sci.*, V, 159—178, (1837) ; *Jour. As. Soc. Beng.*, IV, 530—531, (1835).

Thomson, T.

1. Analysis of a new species of copper ore : *Phil. Trans.* 1814, pp. 45—50.

2. Description and Analysis of a new species of lead ore from India, [Sumatra ?] : *Mem. Wern. Soc. Edin.*, II, 252—258, (1818).

Thomson, T.

1. Western Himalaya and Tibet, a narrative of a journey through the mountains of Northern India during 1847—1848. 8° London, 1852.

2. Sketch of the climate and vegetation of the Himalaya : *Proc. Phil. Soc. Glasgow*, III, 193—204, (1852) ; *Edin. New Phil. Jour.*, 1st series, LII, 309—321, (1852).

Thornton, J.

1. Letter regarding the Coal Beds in the Namsang Naga Hills : *Jour. As. Soc. Beng.*, XVII, 489—491, (1848).

Tickell, S. R.

1. Itinerary, with Memoranda, chiefly Topographical and Zoological, through the Southern portions of the district of Amherst, Province of Tenasserim : *Jour. As. Soc. Bengal*, XXVIII, 421—456, (1859).

Tod, J.

1. Annals and antiquities of Rajasthan, or the Central and Western Rajpoot States of India. 2 vols., 4°, London, 1829—1832 ; *Ditto*, 2 vols., 4°. Calcutta, 1877—1879.

Todd, Major.

1. Report of a journey from Herat to Simla, *viâ* Candahar, Cabool, and the Panjab, undertaken in the year 1838, &c.: *Jour. As. Soc. Beng.*, XIII, 339—360, (1844).

Torrens, H.

1. Note, with a specimen of Iron from the Dhunakar hills, Birbhum : *Jour. As. Soc. Beng.* XIX, 77-78, (1850).

Townsend, R. A.

1. Report on the Petroleum Exploration at Khâtan : *Rec. Geol. Surv. Ind.*, XIX, 204—210, (1886).

Traill, G. W.

1. Account of Hot springs and volcanic appearances in the Himalaya Mountains: *Edin. Jour. Sci.*, VII, 52—56, (1827). .

2. Statistical sketch of Kumaon: *As. Res.*, XVI, 137—234, (1828); KUMAON, pp. 1—49.

Tregear, Vincent.

1. Note on the River Goomtee, with a section of its bed: *Jour. As. Soc. Beng.*, VIII, 712-713, (1839).

Tremenheere, C. W.

1. Description of the mode adopted in taking the observations recorded below to determine the velocity of, and the amount of solid matter in, water at different depths in the *Indus* and in some of the canals in Sind: *Prof. Pap. Ind. Eng.*, 1st series, II, 18—30, (1865).

2. [Memorandum on Mr. Pye Smith's experiments on artificial stone in Bombay]: *Prof. Pap. Ind. Eng.*, 1st series, VII, 271—273, (1870).

3. Notes on the Physical Geography of the Lower Indus: *Brit. Ass. Rep.*, 1866, pt. ii, pp. 177.

Tremenheere, G. B.

1. Letters forwarding a Paper on the formation of the Museum of Economic Geology of India: *Jour. As. Soc. Beng.* IX, 973—996, (1840).

2. Report on the Tin of the Province of Mergui: *Jour. As. Soc. Beng.*, X, 845—851, (1841); XI, 24, 289, (1842); *Cal. Jour. Nat. Hist.*, III, 47—54, (1843); *Sci. Rec. Beng. Gov.*, VI, 5—11, (1852); BURMA, 350—356, (1882); INDO-CHINA, No. 1, I, 251—259, (1886).

3. Report on the Manganese of the Mergui province: *Jour. As. Soc. Beng.* X, 852—853, (1841); *Cal. Jour. Nat. Hist.*, III, 55—56, (1843); *Sel. Rec. Beng. Gov.*, VI, 12—13, (1852); BURMA, 356—357, (1882).

4. Second report on the tin of Mergui: *Jour. As. Soc. Beng.*, XI, 839—851, (1842); INDO-CHINA, I, 260—271, (1886).

5. Report on the Tenasserim coal-field: *Cal. Jour. Nat. Hist.*, II, 417—430, (1842).

6. Report of a visit to the Pakchan River, and of some Tin Localities in the Southern Portion of the Tenasserim Provinces: *Jour. As. Soc. Beng.*, XII, 523—534, (1843); INDO-CHINA, No. 1, I, 275—284, (1886).

7. Report, &c., with information concerning the price of the tin ore of Mergui, *Jour. As. Soc. Beng.* XIV, 329—332, (1845); INDO-CHINA, 298—301 ; (1886).

Tremenheere, G. B., *and* Lemon, *Sir* Charles.

1. Report on the tin of the province of Mergui in Tenasserim, in the northern part of the Malayan Peninsula ; with introductory remarks : *Trans. Geol. Soc. Cornwall*, VI, 68—75, (1846).

Trevor, W. S.

1. Report on the district of Pegu: *Sel. Rec. Gov. Ind.*, XV, 35—45, (1856

Trotter, H.

1. Account of the Pundit's Journey in great Thibet from Leh in Ladakh to Lhása, &c.; *Jour. Roy. Geog. Soc. Lond.*, XLVII, 102—104, (1878), [Gold-fields of Thibet].

Tschermak, G.

1. Kalisalz aus ostendien; *Min. Mitth.*, 1873, p. 135; *Rec. Geol. Surv. Ind.*, VII, 64, (1874).

Turner, S.

1. An account of an embassy to the Court of the Teshoo Lama, in Tibet; containing a narrative of a Journey through Bootàn, and part of Tibet. Also observations, botanical, mineralogical, and medical, by Robert Saunders. 4°. London, 1800.

Tween, A.

1. Memorandum [on the composition of water from the hot springs of Pai, Tavoy district] : *Jour. As. Soc. Beng.*, XXXII, 386, (1863).

2. Analysis of Raniganj coals : *Rec. Geol. Surv. Ind.*, X, 155—158, (1877).

3. Report on the peat of Pertabgurh : *Proc. As. Soc. Beng.*, 1865, p. 87.

Twemlow, G.

1. Flint cores from the Indus : *Geol. Mag.*, IV, 2nd decade, 43, (1867).

Tytler, W. B.

1. The Soan and Koela coal-field : *Jour. As. Soc. Beng.*, VII, 964—965, (1838).

U

Ure, A.

1. A dictionary of arts, manufactures, and mines ; containing a clear exposition of their principles and practice. 8°, London, 1839; 5th edition, 3 vols., London, 1860.

2. Analysis of iron ores from Tavoy and Mergui, and of limestone from Mergui : *Jour As. Soc. Beng.*, XII, 236—239, (1843) ; INDO-CHINA, No. 1, I, 272—275, (1886).

V

Vélain, Charles.

1. Mission de l'île Saint-Paul. Recherches géologiques faites à Aden, à la Réunion, aux îles Saint-Paul et Amsterdam, aux Seychelles. 4.° Paris, 1879.

Verchere, A. M.

1. Notes to accompany a geological map and section of the Lowa Ghur, or Sheen Ghur, range, in the district of Bunnoo, Punjab, with analyses of the Lignites : *Jour. As. Soc. Beng.*, XXXIV, pt. ii, 42—47, (1865).

2. Kashmir, the Western Himálaya and the Afghan Mountains, a geological paper, with a note on the fossils by M. Edouard de Verneuil : *Jour. As. Soc. Beng.*, XXXV, pt. ii, 89—133, 159—203, (1866) ; XXXVI, pt. ii, 9—50, 83—114, 201—209, (1867).

Vertue, J. H. M.

1. General Description of the country between Parvatipore and Jeypore : *Mad. Jour. Lit. Sci.*, XXI, (new series VI,) 264—296, (1859).

Viator.

1. A trip to the Johore river [tin mines] ; Moor's *Indian Archipelago.* 4°, Singapore, 1837, pp. 264—268.

Vicary, M.

1. List of specimens from Bilwan : *Jour. As. Soc. Beng.*, IV, 571, (1835).
2. Geological Report on part of the Beloochistan Hills : *Quart. Jour. Geol. Soc.*, II, 260—267, (1846) ; *Cal. Jour. Nat. Hist.*, VII, 385—392, (1847) ; WESTERN INDIA, 521—527, (1857).
3. Notes on the geological structure of parts of Scinde : *Quart. Jour. Geol. Soc.*, III, 334—349, (1847) ; WESTERN INDIA, 501—517, (1857).
4. [Note on fossil bones at Subathoo] : *Quart. Jour. Geol. Soc.*, III, 349, (1847) ; *Jour. As. Soc. Beng.*, XVI, 1266—1267, (1847).
5. On the geology of the Upper Punjab and Peshawar : *Quart. Jour. Geol. Soc.*, VII, 38—46, (1851).
6. On the Geology of a part of the Himalayan Mountains, near Subathoo : *Quart. Jour. Geol. Soc.*, IX, 70—73, (1853).

Vigne, G. T.

1. Some account of the valley of Kashmir, Ghazni, and Kabul : *Jour. As. Soc. Beng.*, VI, 766—777, (1837).
2. Outline of a route through the Punjaub, Cabool, Cashmere, and into Little Thibet in the years 1834—1838 : *Jour. Roy. Geog. Soc.*, IX, 512—516, (1839).
3. A personal narrative of a visit to Ghuzni, Kabul, and Afghanistan, and of a residence at the Court of Dost Mohamed, with notices of Runjit Sing, Khiva, and the Russian expedition. 8°. London, 1840.
4. Travel in Kashmir, Ladak, Iskardo, the countries adjoining the mountain course of the Indus, and the Himalaya, north of the Punjab. 2 vols. 8°. London, 1842.

Voekov, A.

1. On cotton soil in India. 8° Pamph. St. Petersburg, 1880.

Voysey, H. W.

1. On the diamond mines of Southern India : *As. Res.*, XV, 120—125, (1825) ; *Edin. Jour. Sci.*, VI, 97—104, (1827) ; *Phil. Mag.*, LXVIII, 370—376, (1826) ; *Froriep Notizem*, XVI, 129—138, (1827).
2. On the Building Stones and Mosaic of Akberabad or Agra : *As. Res.*, XV, 429—436, (1825).
3. Description of the Native Manufacture of Steel in Southern India : *Jour. As. Soc. Beng.*, I, 245—247, (1832).
4. On the Geological and Mineralogical structure of the Hills of Sítábald, Nágpur, and its immediate vicinity : *As. Res.*, XVIII, pt. i, 123—127, (1833) ; *Glean. Sci.*, II, 27—28, (1830).
5. On some Petrified Shells found in the Gawilgerh Range of Hills, in April, 1823 : *As. Res.*, XVIII, pt. i, 187—194, (1833) ; *Glean. Sci.*, I, 356—359, (1829) ; *Mad. Jour. Lit. Sci.*, I, 330—342, (1834) ; WESTERN INDIA, 84—88.

Voysey, H. W.,—cont.

6. Reports on the Geology of Hyderabad : *Jour. As. Soc. Beng.*, II, 298—305, 392—405, (1833).

7. Extracts from Dr. Voysey's Private Journal when attached to the Trigonometrical Survey in Southern and Central India : *Jour. As. Soc. Beng.*, XIII, 853—862 ; XIX, 190—212, 269—302, (1850), partly reprinted, in WESTERN INDIA, 48—65, (1857).

Vyse, G. W.

1. Geological notes on the river Indus : *Jour. Roy. As. Soc.*, X, 317—324, (1878).

W

Waagen, W.

1. Abstract of results of examination of the Ammonite fauna of Kutch, with remarks on their distribution among the beds and probable age : *Rec. Geol. Surv. Ind.*, IV, 89—101, (1871).

2. Rough section showing the relations of the rocks near Murree (Mari) : *Rec. Geol. Surv. Ind.*, V, 15—18, (1872).

3. On the occurrence of Ammonites, associated with Ceratites and Goniatites in the Carboniferous deposits of the Salt Range ; *Mem. Geol. Surv. Ind.*, IX, 351—358, (1872).

4. Jurassic Fauna of Kach. The Cephalopoda : *Pal. Indica*, Series ix, I, pts. i—iv, (1873—1875).

5. [Note on the Geology of India] : *Zeits. Deutsch. Geol. Gesel.*, XXVIII, 644—646, (1876) : *Rec. Geol. Surv. Ind.*, X, 98—100, (1877).

6. The Salt Range fossils : *Pal. Indica*, series xiii.
   Vol. I, the Productus limestone group.

   Part. i.   Introduction, Pisces, and Cephalopoda, (1879).

   Part ii.   Gasteropoda and Supplement to pt. i, (1880).

   Part iii.   Pelecypoda, (1881).

   Part iv.   Brachiopoda, (1882—1885).

   Part v.   Bryozoa, Annelida, Echinodermata, (1886).

   Part vi.   Cœlenterata, (1886).

   Part vii.   Cœlenterata, (1888).

7. Uber einige strittige Punkte in der Geologie Indiens : *Neu. Jahrb. Min. Geol.*, 1879, pp. 559—562.

8. Ueber die Geographische Vertheilung der Fossilen organismen in Indien : *Denks. K. K. Akad. Wiss. Wien*, XXXIX, abth. 2, 1—27, (1879) ; *Rec. Geol. Surv. Ind.*, XI, 267—301, (1878).

9. Note on the Attock Slates and their probable geological position : *Rec. Geol. Surv. Ind.*, XII, 183—185, (1879).

10. Uber *Anomia Lawrenciana,* Kon : *Neu Jahrb. Min. Geol.*, 1882, band I, 115—122.

11. On the genus *Richthofenia,* Kags. : *Rec. Geol. Surv. Ind.*, XVI, 12—19, (1883).

Waagen, W.,—cont.

12. Section along the Indus from the Peshawar valley to the Salt Range : *Rec. Geol. Surv. Ind.*, XVII, 118—123, (1884).

13. Notes on some Palæozoic Fossils recently collected by Dr. H. Warth in the Olive group of the Salt Range : *Rec. Geol. Surv. Ind.*, XIX, 22—38, (1886).

14. Die carbone eiszeit : *Jahrb. K. K. Geol. Reichs. Wien*, XXXVII, pt. ii, 143—192, (1887).

Waagen, W. *and* Wynne, A. B.

1. The geology of Mount Sirban in the Upper Punjab : *Mem. Geol. Surv. Ind.*, IX, 331—350, (1872).

Wade, C. M.

1. Notes taken in 1829, relative to the Territory and Government of Iskár-doh, from information given by Charágh Ali, an agent deputed to him in that year by Ahmad Sháh, the Gelp, or ruler, of that country : *Jour. As. Soc. Beng.*, IV, 589—601, (1835).

2. Note on the Hot spring of Lohand Khad : *Jour. As. Soc. Beng.*, VI, 153—154, (1837).

Waldie, D.

1. On Burmese Parafine : *Proc. As. Soc. Beng.*, 1866, pp. 72-73.

2. On Pseudomorphs of Peroxide of Iron after Pyrites : *Proc. As. Soc. Beng.*, 1866, p. 136.

3. Analysis of the Khettree meteorite, with an account of its fall : *Jour. As. Soc. Beng.*, XXXVII, pt. ii, 252, (1869).

4. Analysis of a new Mineral from Burma (O'Rileyite) : *Proc. As. Soc. Beng.*, 1870, pp. 279—283.

Walker, J. T.

1. Memorandum on the earthquake in the Bay of Bengal on 31st December 1881 : *Proc. As. Soc. Beng.*, 1883, pp. 60—62 ; *Rep. Surveyor General Ind.*, 1881—1882, pp. 53—55, (1883).

Wall, P. W.

1. Report on a reputed coal formation at Kota on the Upper Godavery river : *Mad. Jour. Lit. Sci.*, XVIII, (new series, II,) 256—270, (1857)

2. Report on the Lead ores in the Cuddapah District : *Mad. Jour. Lit. Sci.*, XX, (new series, IV), 279—289, (1859).

3. Report on the Silver-lead ores of Kurnool and other portions of the ceded districts : *Mad. Jour. Lit. Sci.*, XX, (new series, IV), 289—299, (1859).

4. Iron-making at Roodrur : *Mad. Jour. Lit. Sci.*, XX, (new series, IV), 299—304, (1859).

Walker, A. M.

1. Report on Productions and Manufactures in the district of Hunnumkoon-dah, in the dominions of H. H. the Nizam of Hyderabad : *Jour. As. Soc. Beng.*, X, 386—392, (1841).

2. On the Geology, &c., &c., of Hunnumkoondah (H. H. the Nizam's Territory) : *Jour. As. Soc. Beng.*, X, 471—475, (1841).

Walker, A. M.,—cont.

3. On the Natural Products about the Pundeelah River, H. H. the Nizam's territory : *Jour As. Soc. Beng.*, X, 509—517, 725—735, (1841).

4. Statistical Report on the Circar of Wurungal : *Mad. Jour. Lit. Sci.*, XV, 219—228, (1849).

5. Statistical Report on the Northern and Eastern Districts of the Soubah of Hydrabad : *Mad. Jour. Lit. Sci.*, XVI, 182—233, (1850).

6. Report on boring for Coal at Kotah, a village 10 or 12 miles from the junction of Wurdah river with the Godavery, in the months of April and May, 1848 : *Mad. Jour. Lit. Sci.*, XVII, (new series, I), 261—265, (1857).

Walker, W.

1. Memoir on the coal found at Kotah, &c:, with a note on the anthracite of Duntimnapilly (H. H. the Nizam's Dominions): *Jour. As. Soc. Beng.*, X, 341—344, (1841).

Walters, H.

1. Coal from Sandoway District : *Jour. As. Soc. Beng.*, II, 263—264, (1833).

Ward, T.

1. Note on the geology of Elephant Hill on the Quedah coast [Malay Peninsula] : *Jour. As. Soc. Beng.*, I, 157—158, (1832).

2. A sketch of the geology of Pulo-Pinang and the neighbouring Islands : *Proc. Geol. Soc. Lond.*, I, 392, (1832).

3. Short sketch of the Geology of Pulo-Pinang, and the neighbouring Islands, with a map and sections : *As. Res.*, XVIII, pt. ii. 149—168, (1833) ; INDO-CHINA, I, 201—211, (1886).

Ward, W. J.

1. Analysis of soils and waters from the Reh lands on the Western Jumna Canal : *Sel. Rec. Gov., N.-W. P.*, II, 202—213, (1868).

Warren, John.

1. An account of the Petrifactions near the village of Treevikera in the Carnatic : *As. Res.*, XI, 1—10, (1810).

2. Observations on the Golden Ore found in the Eastern Provinces of Mysore in the year 1802 : *Jour. As. Soc. Beng.*, III, 463—474, (1834).

Warth, H.

1. Report on the Mayo salt mines, Khewra salt range : *Inland Customs Report*, 1869—70, pp 149—179, (1871).

2. Site of the proposed rock salt shaft at Mount Jogi Tilla : *Inland Customs Report*, 1870-71, pp. 175—177, (1872).

3. The Brine Spring of Kalra near Bukrala on the Grand Trunk Road, from Jhelum to Rawalpindee : *Inland Customs Report*, 1870—71, pp. 177—178, (1872).

4. Preliminary Report on the Salt-bearing Strata in the eastern part of the Salt Range, from the Mayo Salt Mines to the Jogi Tilla : *Inland Customs Report*, 1870-1871, pp. 179—184, (1872).

Warth, H.,—cont.

5. Report on the Salt Mines of the Punjab Salt Range, west of Pind Dadun Khan: *Inland Customs Report*, 1870—1871, pp. 184—213, (1872).

6. Report on the Iron Industry in Chota Bhagal, Kangra district: *Punjab Gazette*, 1873.

7. [Notes' on Rock salt in East India]: *Oesterr. Zeits. Berg. Hutten.*, 1876.

8. Notes on the manufacture of Iron and the future of the Charcoal Iron Industry in India. Flscp. Calcutta, 1881.

9. The Iron Works of Dechouri in Kumaon: *Indian Forester*, VI, 211—218, (1881).

10 Geology of Mussoorie: *Ind. Forester*, X, 113—119, (1884).

11. Coprolites at Mussoorie: *Ind. Forester*, X, 422—423, (1884).

12. A tour in the Salt Range, Panjab; *Ind. Forester*, XIII, 157—163, (1887).

13. Analysis of Phosphatic nodules from the Salt Range, Punjab: *Rec. Geol. Surv. Ind.*, XX, 50, (1887).

14. On the identity of the Olive series in the east, with the speckled sandstone in the west of the Salt Range in the Punjab: *Rec. Geol. Surv. Ind.*, XX, 117—119, (1887).

Warth, H. *and* Wynne, A. B.

1. Extracts from Report on the economic aspect of the Trans-Indus salt region, by Dr. H. Warth, with additions by A. B. Wynne: *Mem. Geol. Surv. Ind.*, XI, 299—330, (1875).

Watson, Dr.

1. Account of two hot springs near Jessore: *Jour. As. Soc. Beng.*, XXV, 225 (1856).

Watson, J. L.

1. Report on the Deposits of Iron ore at Loha Bhur Bhur and Dechouree made on 1st November 1855: *Sel. Rec. Gov. Ind.*, VIII, Supplement, pp. 103—108, (1855).

Watson, T. C.

1. Chirra Punji, and a Detail of some of the favourable circumstances which render it an Advantageous Site for the Erection of an Iron and Steel manufactory on an extensive scale: *Jour. As. Soc. Beng.*, III, 25—32, (1834).

Webb, W. S.

1. Memoir relative to a survey of Kumaon, with some account of the principles upon which it has been conducted: *As. Res.*, XIII, 293—310, (1820).

Weller, J. A.

1. On a trip to the Bulcha and Oonta Dhoora Passes, with an eye sketch: *Jour. As. Soc. Beng.*, XII, 78—102, (1843).

Wells.

1. Wells in the Bhagalpúr District, Behar, &c. [shallow artesian]: *Glean. Sci.*, II, 67, (1830).

2. On the best method of procuring a Plentiful supply of wholesome water in the vicinity of Calcutta: *Glean. Sci.*, II, 237—242, (1830).

Western India.

1. Geological papers on Western India, including Cutch, Sinde and the South-East Coast of Arabia, to which is appended a summary of the Geology of India generally; edited for the Government by Henry Carter. 1 vol., 8°, with folio atlas, Bombay, 1857. Contains A. AYTOUN No. 1; T. L. BELL, No. 1; G. BUIST, No. 8; H. CARTER, Nos. 6, 7, 8, 11, 14; A. T. CHRISTIE, No. 1; J. COPLAND, No. 1; S. COULTHARD, No. 1; F. DANGERFIELD, No. 1; P. M. EGERTON, No. 2; R. ETHERSEY, No. 1; H, FALCONER, No. 8; J. FINNIS, No. 2; A. FLEMMING, No. 4; B. E. FRERE, No. 1; C. W. GRANT, No. 2; S. HISLOP, No. 3; HISLOP and HUNTER, No. 1; R. MURCHISON, No. 1; J. T. NEWBOLD, Nos. 22, 23, 30, 33; R. OWEN, Nos. 2, 3; W. H. SYKES, Nos. 1, 2, 3; M. VICARY, Nos. 2, 3; H. W. VOYSEY, Nos. 5, 7.

White, J. S. D.

1. [Letter regarding coal at Thayet Myo]: *Sel. Rec. Gov. Ind.*, X, 70—78, (1856); BURMA, pp. 173—175, (1882).

Wilcox, R.

1. Memoir of a Survey of Assam and the neighbouring countries, executed in 1825-6-7-8: *As. Res.*, XVII, 314—469, (1832).

Wilkins, H. St. C.

1. Extract from a report on attempts made to supply Aden with water: *Jour. Bo. As. Soc.*, V, 597—611, (1857).

Wilkinson, C. J.

1. Sketch of the Geological structure of Southern Konkan: *Rec. Geol. Surv. Ind.*, IV, 44—47, (1871).

Wilkinson, T.

1. [Minerals of Nagpore]: *Cal. Jour. Nat. Hist.*, III, 290—292, (1843).

Williams, C.

1. Extract from Journal of a Trip to Bhamo: *Jour. As. Soc. Beng.*, XXXIII, 189—195, (1864).

Williams, E. C. S.

1. Pegu: its Geography, descriptive and physical: *Sel. Rec. Gov. Ind.*, XX, 4—5, (1856).

Williams, H.

1. Eruption of a submarine volcano seen from Kyouk Phyu: *Jour. As. Soc. Beng.*, XIV, p. xxiv, (1845).

2. A Geological Report on the Damoodah valley. 8°. London, 1850.

3. Geological Report on the Kymore Mountains, the Ramghur coal-fields, and on the manufacture of Iron, &c. 8°. Calcutta, 1852.

Wilson, W. L.

1. [Stone implements from Saugor district]: *Proc. As. Soc. Beng.*, 1867, p. 142.

Winchester, J. W.

1. Note on the Island of Karrack in the Gulf of Persia: *Trans. Bo. Geog. Soc.*, Nov. 1838, pp. 35—39.

Wislicenus, J.

1. Mineralanalysen. I, Rothkupfererz von Landu in Bengalen: *Vierteljahrs. Naturf Ges. Zurich,* VII, 17—20, (1862); *Zeits Gesammt. Naturw. Halle,* XX, 196—198, (1862).

2. Mineral Analysen; Magneteisenstein von Landu in Bengalen: *Vierteljahrs. Naturf. Ges. Zurich,* VII, 258—264, (1862); *Zeits. Gesammt. Naturw. Halle,* XX, 198—201, (1862)

Wojeikoff.

1. A Russian account of scientific Progress in India [Black cotton soil]: *Nature,* XVI, 425, (1877); [Quoting *Investia Imp. Russ. Geog. Soc.*]

Wood, Browne.

1. Extracts from a report of a journey into the Naga Hills: *Jour. As. Soc. Beng.,* XIII, 771—785, (1844).

Wood, J.

1. Report on the River Indus (Sections 1—5): *As. Soc. Beng.,* X, 518—569, (1841); *Sel. Rec. Gov. Bombay,* XVII, 541—588, (1855).

2. A personal narrative of a journey to the source of the river Oxus by the route of the Indus, Kabul and Badakshan, performed under the sanction of the Supreme Government of India in the years 1836, 1837, and 1838. 8°. London, 1841; new edition, 1872.

Wood, Martin.

1. [Gold-mining in India]: *Jour. Soc. Arts,* XXX, 592, (1882).

Woodburn.

1. The Sambur Salt Lake: *Trans. Bo. Geog. Soc.,* XII, Appendix, 15—17, (1856).

Woodward, S. P.

1. Note on Land and Freshwater shells collected by Capt. Godwin-Austen [in Cashmere territory]: *Quart. Jour. Geol. Soc.,* XX, 388, (1864).

Wragge, R.

1. On peat and its profitable utilisation in Indian Locomotives, and for other purposes. [Peat in the Neilgherries]: *Jour. Soc. Arts,* XIX, 201—208, (1871).

Wright, B. W.

1. A short notice of Earthquakes: *Mad. Jour. Lit. Sci.,* I, 104—111, (1834).

Wynne, A. B.

1. On the Geology of the Island of Bombay: *Mem. Geol. Surv. Ind.,* V, pt. ii, 173—225, (1866).

2. On Denudation and its causes [slight references to Indian Physical Geology]: *Geol. Mag.,* 1st Decade, IV, 3—11, (1867).

3. Remains of Prehistoric Man in Central India [Aurangabad]: *Geol. Mag.,* 1st decade, III, 283, (1866).

4. Geological notes on the Surat Collectorate: *Rec. Geol. Surv. Ind.,* I, 27—32, (1868); *Bombay Gazetteer,* II, 29—36, (1877).

Wynne, A. B.,—cont.

5. Notes on some physical features of the Land formed by Denudation: *Jour. Roy. Geol. Soc. Dublin*, new series, I, 256—267, (1868).

6. The valley of the Poorna River, West Berar: *Rec. Geol. Surv. Ind.*, II, 1—5, (1869).

7. Preliminary notes on the Geology of Kutch, Western India: *Rec. Geol. Surv. Ind.*, II, 51—59, (1869).

8. On the occurrence of frog-beds at a locality hitherto concealed, but exposed now by reclamation works in Bombay Island, December 1867: *Mem. Geol. Surv. Ind.*, VI, 385—386, (1869).

9. On the petroleum locality of Sudkal, near Futtijung, West of Rawal-pindi: *Rec. Geol. Surv. Ind.*, III, 73—74, (1870).

10. On the Geology of Mount Tilla, in the Punjab: *Geol. Surv. Ind.*, III, 81—86, (1870).

11. Memoir on the Geology of Kutch, to accompany a map compiled by A. B. Wynne and F. Fedden, during the seasons 1867—1868 and 1868—1869 : *Mem. Geol. Surv. Ind.*, IX, pt i, 1—289, (1872).

12. Notes·from a progress report on the geology of parts of the Upper Panjáb : *Rec. Geol. Surv. Ina.*, VI, 59—64, (1873).

13. Observations on some features in the Physical Geology of the outer Himalayan region of the Upper Panjáb, India: *Quart. Jour. Geol. Soc.*, XXX, 61—80, (1874).

14. Notes on the Geology of Mari hill station in the Panjáb: *Rec. Geol. Surv. Ind.*, VII, 64—74, (1874).

15. The Trans-Indus Salt Region in the Kohat district: *Mem. Geol. Surv. Ind.*, XI, pt. ii, 105—298, (1875).

16. Geological notes on the Khárian hills in the Upper Punjab: *Rec. Geol. Surv. Ind.*, VIII, 46—48, (1875).

17. Note on the tertiary zone and underlying rocks in the North-Western Panjáb : *Rec. Geol. Surv. Ind.*, X, 107—132, (1877).

18. On "Remarks, &c." by Mr. Theobald upon Erratics in the Panjáb : *Rec· Geol. Surv. Ind.*, XI, 150-151, (1878).

19. On the Geology of the Salt Range in the Punjab: *Mem. Geol. Surv. Ind.*, XIV, (1878).

20. Notes on the Physical Geology of the Punjab: *Quart. Jour. Geol. Soc.*, XXXIV, 347—375, (1878).

21. Notes on the Earthquake in the Punjab of March 2nd, 1878: *Jour. As. Soc. Beng.*, XLVII, pt. ii, 131—140, (1878).

22. A geological reconnaissance from the Indus at Kushalgarh to the Kurram at Thal on the Afghan frontier : *Rec. Geol. Surv. Ind.*, XII, 100—114, (1879).

23. Further notes on the Geology of the Upper Panjáb: *Rec. Geol. Surv. Ind.*, XII, 114—133, (1879).

24. On the continuation of the road section from Murree to Abbottabad : *Rec. Geol. Surv. Ind.*, XII, 208—210, (1879).

25. Recent Publications of the Geological Survey of India: [corrects mis-statements of his views in the Manual of the Geology of India and Palæontologia Indica]: *Geol. Mag.*, 2nd decade, VI 429—431, (1879) ; VII, 48, (1880).

Wynne, A. B.,—cont.

26. On the Trans-Indus Extension of the Punjab Salt Range: *Mem. Geol. Surv. Ind.*, XVII, pt. ii, 211—305, (1880).

27. Travelled blocks of the Punjab: *Rec. Geol. Surv. Ind.*, XIV, 153—154, (1881).

28. On the connection between Travelled Blocks in the Upper Panjáb and a supposed glacial period in India: *Geol. Mag.*, 2nd decade, VIII, 97—99, (1881).

29. Further note on the connection between the Hazara and Cashmir Series: *Rec. Geol. Surv. Ind.*, XV, 164—169, (1882).

30. Notes on some recent discoveries of interest in the geology of the Punjab Salt Range: *Proc. Roy. Dublin Soc.*, 85—93, (1886): *Geol. Mag.*, 3rd decade, III, 131—134, (1886).

31. A facetted and striated Pebble from the Salt Range, Panjab, India: *Geol. Mag.*, 3rd decade, III, 492—494, (1886).

32. On a certain fossiliferous pebble band in the "Olive group," of the Eastern Salt Range, Panjab: *Quart. Jour. Geol. Soc.*, XLII, 341—350, (1886).

33. Phosphatic nodules from the Salt Range: *Geol. Mag.*, 3rd decade, IV, 95, (1887).

Wynne A. B., *and* **Waagen, W.**

1. The Geology of Mount Sirban in the Upper Punjab: *Mem. Geol. Surv. Ind.*, IX, 331—350, (1872).

Wynne, A. B., *and* **Warth, H.**

1. Extracts from Report on the economic aspect of the Trans-Indus' Salt Region, by Dr. H. Warth, with additions by A. B. Wynne: *Mem. Geol. Surv. Ind.*, XI, 299—330, (1875).

Y

Yarkand.

1. Scientific results of the second Yarkand Mission, based upon the collections and notes of the late Ferdinand Stoliczka. Geology, by W. T. Blanford. 4°. Calcutta, 1878.

2. Syringo-sphæridæ, by P. Martin Duncan. 4°. 1879.

Young, C. B.

1. A few remarks on the subject of the Laterite found near Rangoon: *Jour. As. Soc. Beng.*, XXII, 196—201, (1853).

Young, S. D.

1. An account of the General and Medical Topography of the Neilgerries: *Trans. Med. Phys. Soc. Calcutta*, IV, 36—78, (1829).

Yule, H.

1. Notes on the Iron of the Kasia Hills, for the Museum of Economic Geology: *Jour. As. Soc. Beng.*, XI, 853—856, (1842).

Yule, H.,—cont.
2. A narrative of the mission sent by the Governor General of India to the Court of Ava in 1855, with notices of the country, government, and people. General report by Capt. H. Yule; Geological Report, by Dr. T. Oldham. 4°. London, 1858.

Z

1. **Zigno, Achille de.** *See* DE ZIGNO.

Government of India Central Printing Office, No. 3 D. G. Survey.—29-10-88.—505.

Printed in the United States
By Bookmasters